Microalgae Biotechnology for Wastewater Treatment, Resource Recovery and Biofuels

Monika Prakash Rai · Shrasti Vasistha

Microalgae Biotechnology for Wastewater Treatment, Resource Recovery and Biofuels

Towards Sustainable Biorefinery

Springer

Monika Prakash Rai
Amity Institute of Biotechnology
Amity University Uttar Pradesh
Noida, Uttar Pradesh, India

Shrasti Vasistha
Amity Institute of Biotechnology
Amity University Uttar Pradesh
Noida, Uttar Pradesh, India

ISBN 978-3-031-31673-9 ISBN 978-3-031-31674-6 (eBook)
https://doi.org/10.1007/978-3-031-31674-6

This Springer imprint is published by the registered company Springer Nature Switzerland AG
The registered company address is: Gewerbestrasse 11, 6330 Cham, Switzerland

Contents

Chapter 1
Introduction

Increased industrialization and overexploitation of natural resources have affected the environment adversely. Textile, municipal, dairy, pharmaceutical, swine, and aquaculture sectors regularly produce various forms of wastewater (El-Sheekh et al., 2023). Wastewater is high in nutrients, promoting eutrophication and posing a threat to the ecosystem's flora and fauna. According to a report, eutrophication generates an annual loss of over 2 billion dollars and continues to impair real estate and fishing industries. Wastewater comprises various undesired chemical elements and microorganisms that have both acute and long-term consequences for the environment and human health (Sathya et al., 2022). Untreated wastewater can cause gastroenteritis, skin infections, hepatitis A, and a disease similar to leptospirosis (Pratap et al., 2023).

Additionally, the use of untreated wastewater in irrigation has also been linked to a number of issues, including excessive vegetative growth, irregular fruit development, and worse quality and yield. The primary goal of wastewater treatment has traditionally been safeguarding downstream consumers from health hazards (Tauqeer et al., 2022). Generally, wastewater is treated with a combination of physical (screening, grit chambers, floatation, etc.) and chemical (neutralization, flocculation, oxidation) processes. The various chemical processes for wastewater treatment are expensive, frequently resulting in sludge creation and an increased danger of secondary water pollution. Other issues with wastewater treatment include high energy consumption (2–4% of national electric power consumption) and a shortage of efficient and productive workers, wastewater treatment plants have an environmental impact, and introducing new and efficient technology requires expensive steps and replacement of the old infrastructure (Zhu et al., 2023). An additional intended goal, resource recovery from wastewater combined with wastewater treatment technologies, has gained traction in recent decades. Studies reported the production of polyhydroxyalkanoates (PHAs) from various wastewater treatment facilities and proposed that European wastewater treatment plants may provide roughly 120% of global biopolymer output (2016). Wastewater has also been investigated to produce various value-added products such as single-cell protein and carotenoids (Saejung et al., 2020).

M. P. Rai and S. Vasistha, *Microalgae Biotechnology for Wastewater Treatment, Resource Recovery and Biofuels*, https://doi.org/10.1007/978-3-031-31674-6_1

Another crisis is the demand for global energy, mainly fossil fuels, coal, and petroleum which has become an essential factor for economic development (Owusu and Asumadu-Sarkodie, 2016). Yin et al. (2020) reported that energy consumption will increase from 550 EJ to 865 EJ by 2040. The rapid depletion of fossil fuel resources has raised many environmental concerns, including climate change, CO_2 levels, and ecological degradation. In addition, it also provoked an interest in producing sustainable energy sources such as biofuels. Biofuels are a promising approach to lowering fossil fuel dependency and reducing greenhouse emissions (Malla et al., 2023). The development of third-generation feedstock emerged as a potential alternative to conventional energy sources, and less dependency on fossil fuels will help maintain healthy and economic sustainability.

Oxygenated photosynthetic microalgae are prokaryotic or eukaryotic microorganisms in both freshwater and marine habitat (Zhu et al., 2023). Microalgae offer various inherent advantages as they are capable of accumulating lipids (10–70%), capable of the production of pigments, carbohydrate, and proteins, have a high growth rate, possess a high photosynthetic rate, utilize solar energy and CO_2, and convert into potential biomass; their robust nature makes them versatile (Fig. 1.1). The average biofuel production from microalgae is 20–30% more than the vegetable oils and seeds microalgae (Zhu et al., 2023).

Microalgae use in wastewater treatment is helpful since it aids in waste management and generates bioenergy (Karray et al., 2022). Many studies have suggested that using microalgal technology in wastewater treatment could lower wastewater nutrient concentrations by up to 95%. Microalgae may accumulate lipids, produce biohydrogen, and generate bioelectricity in addition to removing nutrients from wastewater (Zhang et al., 2021). Algal biomass is used as an alternative oil source for biodiesel production due to its high lipid content. They can be converted to biodiesel by employing various acids, alkalis, and enzymatic catalysts via transesterification reactions.

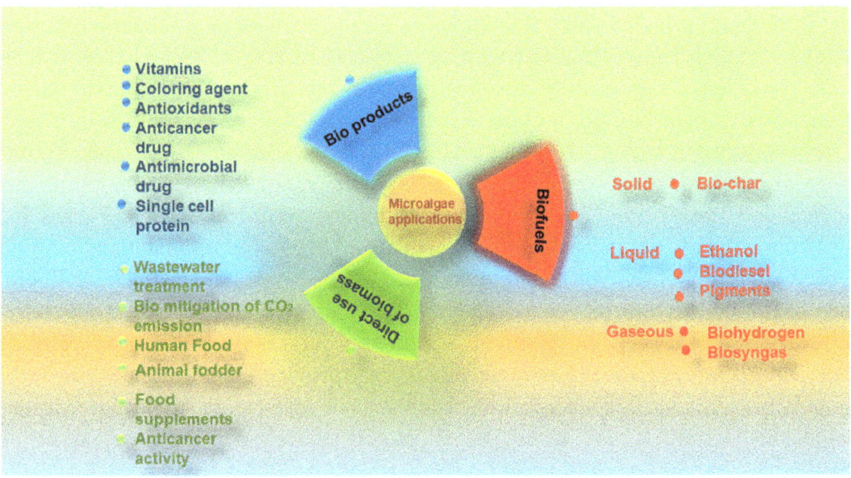

Fig. 1.1 Microalgae applications in various fields

Algal biomass can be used to make biohydrogen using modified procedures such as growing algae in wastewater under aerobic circumstances before moving them to an anaerobic reactor (Bhatia et al., 2021). Fresh algal biomass or residual biomass after lipid extraction can be used immediately for bioenergy production by dark fermentation (biohydrogen), fermentation (bio alcohols), or anaerobic digestion (methane). Microalgae have also been used to generate bioelectricity using a method known as photosynthetic microbial fuel cells (PhotoMFC) (Vinayak et al., 2021). The biodiesel produced from microalgae is biodegradable and less harmful to the environment.

Microalgal cultivation methods have been developed that are both traditional (open and closed systems) and advanced (turf scrubber and hybrid) (Bhatia et al., 2021). Wastewater-based algal biomass is highly affected by selecting the accurate method for cultivation. Separating algal biomass from wastewater is the next stage after successful cultivation. Microalgae harvesting has been documented using a variety of physical, chemical, biological, and electromagnetic approaches (Singh et al., 2014). However, there are a few roadblocks in the way to the widespread use of this technology. Because algae have a negative charge on their surface, which prevents them from producing big particles, harvesting algal biomass is a cost-effective technique (Vasistha et al., 2021a, 2021b). Algal biomass harvesting is predicted to account for roughly 30% of total production expenses. Algal biomass can also be used to make several bioproducts, including nutraceuticals, medicines, and pigments. As a result, combining wastewater treatment with algae and bioenergy generation with other valuable byproducts could be beneficial (Zhang et al., 2021). Despite these potential characteristics and different applications of microalgae, its commercial scale is challenging. The high cultivation cost and low lipid yield are significant bottlenecks in advancing microalgae-based technology (Raj et al., 2022). Additionally, the biorefinery approach is a promising concept for producing cost-effective biomass; as aforementioned, the microalgae mass is converted into multiple products (Krishnamoorthy et al., 2022).

This book is designed to study microalgae and their application in the broadest sense. The book section comprehensively covers the advanced microalgae cultivation techniques, including three-chamber microalgae-bio-electrochemical system designed for clean microalgae cultivation, microalgae bacterium membrane PBR, and BIO-ALGAE simulation tool. Wastewater remediation and nutrient recovery by microalgae, lipid extraction for biodiesel, and utilization of residual biomass for different applications. They are further focusing on different industries and research organizations involved in microalgae-based technology for global marketing, for example, Sunchlorella, Cellana, algosource, etc., which is the most significant feature of this book.

Moreover, significant inconsistencies which have been explored in microalgae technology are microalgae cultivation, lipid extraction, and further utilization of residual biomass, which affect the economic viability. Acknowledging these challenges, the limitation, outcome, and economic aspects are discussed. Hence, this book summarizes the current state of microalgae-based biorefinery techniques to establish a long-term and cost-effective process.

References

Bhatia, S. K., Mehariya, S., Bhatia, R. K., Kumar, M., Pugazhendhi, A., Awasthi, M. K., Atabani, A. E., Kumar, G., Kim, W., Seo, S.-O., & Yang, Y.-H. (2021). Wastewater based microalgal biorefinery for bioenergy production: Progress and challenges. *Science of the Total Environment, 751*, 141599.

El-Sheekh, M. M., Galal, H. R., Mousa, A. S. H., & Farghl, A. A. (2023). Coupling wastewater treatment, biomass, lipids, and biodiesel production of some green microalgae. *Environmental Science and Pollution Research*, S, 1–13.

Karray, R., Elloumi, W., Ali, R. B., Loukil, S., Chamkha, M., Karray, F., & Sayadi, S. (2022). A novel bioprocess combining anaerobic co-digestion followed by ultra-filtration and microalgae culture for optimal olive mill wastewater treatment. *Journal of Environmental Management, 303*, 114188.

Krishnamoorthy, A., Rodriguez, C., & Durrant, A. (2022). Sustainable approaches to microalgal pre-treatment techniques for biodiesel production: A review. *Sustainability, 14*(16), 9953.

Malla, F.A., Bandh, S.A., Wani, S.A., Hoang, A.T. and Sofi, N.A., 2023. Biofuels: Potential Alternatives to Fossil Fuels. In *Biofuels in Circular Economy* (pp. 1–15). Singapore: Springer Nature Singapore.

Owusu, P.A. and Asumadu-Sarkodie, S. (2016). A review of renewable energy sources, sustainability issues and climate change mitigation. Cogent Engineering, *3*(1), 1167990.

Pratap, B., Kumar, S., Nand, S., Azad, I., Bharagava, R. N., Ferreira, L. F. R., & Dutta, V. (2023). Wastewater generation and treatment by various eco-friendly technologies: Possible health hazards and further reuse for environmental safety. *Chemosphere, 313*, 137547.

Raj, T., Morya, R., Chandrasekhar, K., Kumar, D., Soam, S., Kumar, R., Patel, A.K., & Kim, S. H. (2022). Microalgae biomass deconstruction using green solvents: Challenges and future opportunities. *Bioresource Technology*, S. 128429.

Saejung, C., & Salasook, P. (2020). Recycling of sugar industry wastewater for single-cell protein production with supplemental carotenoids. *Environmental Technology, 41*(1), 59–70.

Sathya, K., Nagarajan, K., Malar, C. G. G., Rajalakshmi, S., & Lakshmi, P. R. (2022). A comprehensive review on comparison among effluent treatment methods and modern methods of treatment of industrial wastewater effluent from different sources. *Applied Water Science, 12*(4), 70.

Singh, B., Guldhe, A., Rawat, I., & Bux, F. (2014). Towards a sustainable approach for development of biodiesel from plant and microalgae. *Renewable and Sustainable Energy Reviews, 29*, 216–245.

Tauqeer, H. M., Turan, V., & Iqbal, M. (2022). Production of safer vegetables from heavy metals contaminated soils: the current situation, concerns associated with human health and novel management strategies. *Advances in bioremediation and phytoremediation for sustainable soil management: Principles, monitoring and remediation* (pp. 301–312). Springer International Publishing.

Vasistha, S., Khanra, A., Clifford, M., & Rai, M. P. (2021a). Current advances in microalgae harvesting and lipid extraction processes for improved biodiesel production: A review. *Renewable and Sustainable Energy Reviews, 137*, 110498.

Vasistha, S., Khanra, A., & Rai, M. P. (2021b). Influence of microalgae-ZnO nanoparticle association on sewage wastewater towards efficient nutrient removal and improved biodiesel application: An integrated approach. *Journal of Water Process Engineering, 39*, 101711.

Vinayak, V., Khan, M. J., Varjani, S., Saratale, G. D., Saratale, R. G., & Bhatia, S. K. (2021). Microbial fuel cells for remediation of environmental pollutants and value addition: Special focus on coupling diatom microbial fuel cells with photocatalytic and photoelectric fuel cells. *Journal of Biotechnology, 338*, 5–19.

Yin, Z., Zhu, L., Li, S., Hu, T., Chu, R., Mo, F., Hu, D., Liu, C., Li, B. (2020). A comprehensive review on cultivation and harvesting of microalgae for biodiesel production: Environmental pollution control and future directions, *Bioresource Technology, 301*, 122804. https://doi.org/10.1016/j.biortech.2020.122804.

Zhang, C., Li, S., & Ho, S. H. (2021). Converting nitrogen and phosphorus wastewater into bioenergy using microalgae-bacteria consortia: A critical review. *Bioresource Technology, 342*, 126056.

Zhu, S., Higa, L., Barela, A., Lee, C., Chen, Y., & Du, Z. Y. (2023). Microalgal consortia for waste treatment and valuable bioproducts. *Energies, 16*(2), 884.

Chapter 2
Wastewater-Mediated Microalgae Isolation and Culturing Techniques

Microalgae are a diverse group of organisms that can be found in various environments. They differ according to their habitats and metabolite synthesis. Microalgae isolation and purification are critical steps in their biotechnological uses. The goal of algae isolation and purification is to obtain axenic cultures or viable cultures of a single species free of contaminants (Singh et al., 2015). Any contaminant can have a significant financial impact, result in the loss of a batch, or drastically alter the final product's composition.

2.1 Pre-isolation Techniques

Prior to isolation, several procedures are performed; attaining the axenic cultures will be aided by the collection, proliferation, and separation into smaller groups.

2.1.1 Sample Collection From Wastewater

The collection of axenic culture is a critical step of the process. Failure occurs when algal cells are damaged during harvesting. While sampling an unknown environment or target organism, multiple collection methods should be considered. Syringe sampling, scraping, brushing, and the inverted Petri dish method are examples of sample techniques (Brenan and Owende, 2010). However, the most commonly reported sample method is simply collecting water from a body of water. A Niskin bottle or rosette sampler can be used if the target species is generally located in depth. When sampling, it is fundamental to record factors such as light, water temperature, dissolved O_2 and CO_2, nutrient concentration, pH, and salinity to mimic these conditions at the laboratory and establish the location coordinates to replicate sampling if needed (Singh et al., 2015).

M. P. Rai and S. Vasistha, *Microalgae Biotechnology for Wastewater Treatment, Resource Recovery and Biofuels*, https://doi.org/10.1007/978-3-031-31674-6_2

2.1.2 Enrichment of Sample

Once a sample has been transferred to the lab, nutrients can be added to the natural samples to boost microalgal growth. Because organic chemicals might cause unwanted bacterial multiplication, they should only be used in limited doses. Individual cells are separated from the enriched culture after being incubated and tested for the growth of the target species every few days (Mutanda et al., 2011). The target species may grow efficiently after isolation but perish after many transfers, signifying that the culture medium is deficient in an element or compound or that the organism has accumulated wastes that poison the environment.

Enrichment of samples should be targeted for different microalgae; for example, enriching samples with ES-enriched saltwater may benefit fragile algae and flagellates; chrysophytes, chlorophytes, diatoms, and cryptophytes grow well when f/2-medium is added to water samples (Fernandez et al., 2021). On the other hand, cyanobacteria growth is aided by the addition of the BG11 medium.

2.1.3 Filtration

The filtration technique governs the separation of samples into two groups based on particle size differences. The microalgae (mostly coccoid) were isolated from Moroccan Mediterranean saltwater by filtering samples through a series of membranes with decreasing mesh (33, 20, and 0.45 m) and then placing membranes on agar plates (Fernandez et al., 2021). This approach allows microorganisms to be separated solely by their cell size, but it is usually insufficient to generate axenic or unialgal cultures because various organisms can have identical cell sizes (Singh et al., 2015).

2.1.4 Centrifugation

Centrifugation uses gravity to separate organisms by cell size. Low-density cells, such as bacteria, tend to remain in the supernatant fraction and are discarded, whereas microalgae, as higher-density organisms, are typically found in the pellet fraction (Price, 2014). After numerous centrifuging cycles, a higher concentration of a desired microalgal strain will be produced. Separating the individual cells into different bands can be done by a silica or Percoll density gradient. Centrifugation separation effectiveness is determined by speed and time; thus, these parameters should be tailored for a specific species. Excessive centrifugal force has also been shown to harm delicate cells. Researchers used a 90%, 90–50%, 90–50–30 %, and 90–50–30–10% Percoll solution to density gradient centrifugate the dinoflagellate species Karenia mikimotoi and Alexandrium tamarense, harvesting the algal cells and reducing the bacterial population from 5.79 0.22 log10 CFU/mL to 1.13 0.07 log10 CFU/mL (Lee et al., 2021a, 2021b; Fernandez et al., 2021). Centrifugation has been shown in various studies to be effective in concentrating algae cells and lowering the bacterial population. This procedure, however, does not produce an axenic algal culture on its own; rather, it facilitates subsequent techniques that ensure the purity of cultures.

2.2 Techniques for Microalgae Isolation From Wastewater

2.2.1 Single-Cell Isolation

A micropipette, a Pasteur pipette, or a glass capillary can be used for this procedure. The goal is to take a single cell from the sample, place it in a sterile droplet of culture medium, then pick, and place it in another sterile droplet (Andersen & Kawachi, 2005). This procedure is continued until just one algal cell remains in the final isolation vessel. Micromanipulation using a stereomicroscope, microcapillary tubes, and optical tweezers is a novel technology for single-cell separation that dramatically lowers labor and cell damage (Singh et al., 2015). They do, however, necessitate costly equipment and highly trained lab employees. These approaches are helpful for screening and isolation because they perform cell separation with high precision. Microcapillaries are typically composed of glass and have a 1 mm outside diameter.

On the other hand, optical tweezers (OT) are a non-invasive method for physical manipulation that uses a highly focused laser beam to trap and manipulate microscopic objects, typically in the infrared (IR) region (wavelengths between 700 and 1300 nm). It generates two forces: the scattering force produced by photons striking the cell along their propagation direction and the gradient force produced by a gradient of field interference (Fernandez et al., 2021). These forces work as optical tweezers and are dependent on the laser beam wavelength and particle size. This enables the micromanipulation of a single item with a size ranging from nanometers to micrometers, including living cells. OT for single-cell manipulation can be based on fluorescence activation or differences in cell shape, size, and refractive index, allowing cells to be sorted into compartments without any pre-treatment. However, there have been very few investigations on this specific technique. Pilat et al., determined that laser wavelengths longer than 935 nm might be used to perform non-invasive optical manipulation of living cells of photosynthetic microorganisms. They also discovered that a wavelength of 1064 nm did not harm the microalgae and is sufficient for reliable 3D cell manipulation. To obtain pure cultures of *Nannochloropsis* sp., *Dunaliella salina*, *Chlorella* sp., and *Tetraselmis* sp., the serial dilution and posterior micromanipulator technique (Cadoret et al., 2014).

2.2.2 Isolation in Agar

The algae must be able to grow on solid media in order to be isolated successfully using this procedure. Coccoid cells and most diatoms thrive on agar, although certain flagellates and dinoflagellates do not. Following agar culturing, the selected cells are picked up using a micropipette tip and discharged into a liquid culture media (Fernandez et al., 2021). Perpetual transfer with frequent use of the agar pour technique can also be utilized to maintain axenic cultures. In addition, the isolation of microalgae is performed by filtering samples through a 100 m filter, enriching the samples with f/2-medium, streaking the enriched microalgal communities on f/2-medium solidified plates with agar, picking colonies from the plates, and transferring them to a liquid f/2-medium in 12-well plates (Jhan et al., 2014). Various culture media are used to isolate microalgae cells such as Thao, identified oleaginous microalgae by using BG11 for fresh water, f/2 for brackish water, and f/2 for marine water samples after sampling. After that, a solid or semi-solid medium was made for each of the three media by adding 10 g/L or 5 g/L agarose, respectively (Thao et al., 2017). Isolation was accomplished using two methods:

(1) streaking on solid agarose plates;
(2) Mixing 0.1 mL of enriched cultures with 5 mL of semi-solid medium, vortexed, and pouring over the solid medium to form a semi-solid overlay, incubating, and repeating until pure isolates were obtained. In order to attain the best result microalgae, properties like phototaxis can be associated with agar cultivation. *Chlamydomonas* has flagella, and the rhodopsin pigment is attracted to light. Isolation of axenic culture in agar medium benefits only if the targeted microalgae to be isolated do not grow on consortia with other species (Banerjee et al., 2012).

Additional testing and microscopic observation may be required to ensure the culture's purity. Other approaches should be used if the desired microalgae remain contaminated after isolation in agar. On the other hand, this technique allows for the isolation of specific microalgae from the majority of microorganisms in a sample, resulting in a significant reduction in pollutants. Figure 2.1 illustrates the various isolation techniques.

2.2.3 Dilution Techniques

The technique is diluting a sample or culture and putting only one cell into a vessel (Andersen, 2005). Preparing and filling test tubes with 9 mL of sterile saline solution or enough culture media is typical for this procedure. Then, in the first test tube, 1 mL of microalgae culture or an enhanced sample is transferred and stirred to create a 1:10 dilution. After that, 1 mL of the first dilution is transferred to the second tube, and the process is repeated five to six times. Because some single cells will

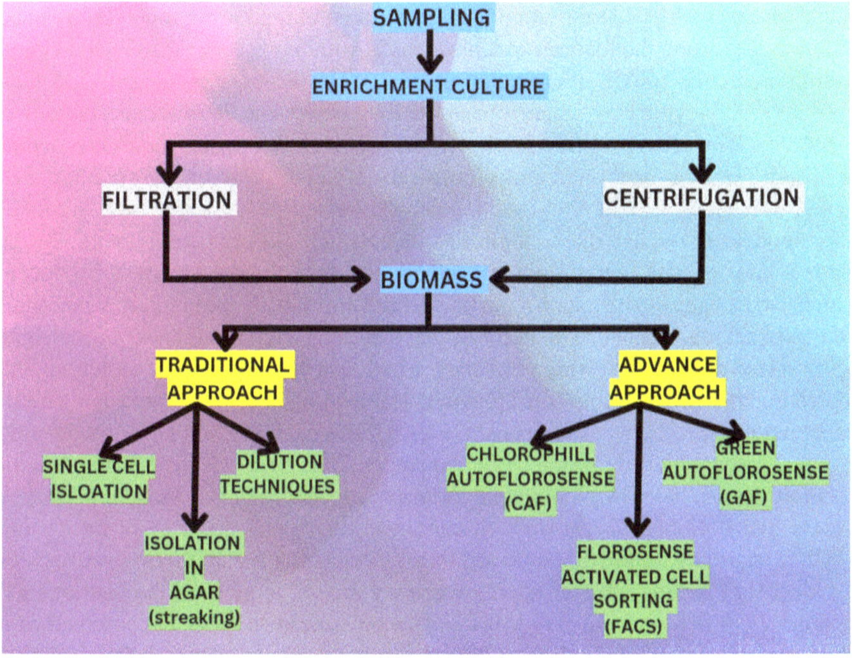

Fig. 2.1 Microalgae isolation traditional and advanced approach

perish and some inocula will include two or more species or no cells, it is crucial to consider inoculating numerous tubes from the last dilution. Enrichment substrates can be added to isolation tubes to grow or incubate a target species under precise conditions (Fernandez et al., 2021). The dilution technique is rarely used to create axenic isolates since bacteria are usually more prevalent than algae. When attempting to culture random algal species from field samples to identify and investigate novel species, the dilution approach is frequently used. Dilution procedures allow for the creation of cultures that are largely made up of a few microalgal species, but they do not guarantee culture purity (Alam et al., 2019). The contamination tests and microscope observation are required to detect additional bacteria in the culture. In order to create the axenic microalgal culture, further procedures will most likely be required.

2.3 Advanced Techniques for Microalgae Isolation in Wastewater

Flow cytometry is the most widely used automated technology for microalgae cell sorting because of the autofluorescence of their natural internal colors, which allows for the differentiation of various microalgae species from other microorganisms.

Algae species such as diatoms, dinoflagellates, and prokaryotic phytoplankton can be differentiated using the features of Chlorophyll Autofluorescence (CAF) and Green Autofluorescence (GAF) (Fernandez et al., 2021). Fluorescence-Activated Cell Sorting (FACS) cytometry, which is based on light scatter and fluorescence, is gaining popularity. It involves passing a fluid stream through a laser beam, where cells scatter or absorb the laser beam and emit fluorescence, which provides information such as cell size, integrity, and photosynthetic characteristics. Microalgae cells are naturally fluorescence-active due to their high chlorophyll content (Duong et al., 2012). After a laser excites chlorophyll, an intense red fluorescence signal is produced, which can be used as a trigger for automated cell sorting, allowing microalgae isolation and bacteria removal. The number, cell size, shape, and toughness of the algae being sorted determine sorting efficiency, which can be improved by enriching and sonicating the culture before sorting. After establishing unialgal cultures using traditional methods, FACS is used to create axenic cultures, for which DNA sequencing revealed no bacterial ribosomal DNA sequences (Bui, 2013). *Scenedesmus, Actinotaenium, Cosmarium, Haplotaenium, Staurastrum, Aulacoseira, Navicula, Thalassiosira*, and *Protoperidinium* were among the microalgae isolated by Fistarol using FACS (Fistarol et al., 2018). However, they reported that not all microalgae species are ideal for FACS sorting, either because they are too large to fit the instruments' capillary lines or too sensitive to withstand the pressures used. FACS, combined with fluorescent dyes such as Bodipy or Nile red, is an excellent approach for screening the production of metabolites of interest by organisms (Pereria et al., 2013).

2.3.1 *Identification of Microalgae Strains Using Artificial Intelligence*

The emerging technology of Artificial Intelligence (AI) can be applied in major three areas of microalgae technology. The first area, which combined AI with microscope image processing, produced a quick way for identifying species. Another area focuses to evaluate cell concentration in microalgal culture, combined AI technologies with fluorescence measurement-based techniques (Teng et al., 2020) while the third section concentrated on using AI and a spectrometer based on matrix-assisted laser desorption (MALDI) to characterize microalgae from their mass spectra (Andrade et al., 2015; Barbano et al., 2015). The development of species identification was primarily aimed at ecological monitoring of aquatic habitats, such as tracking cyanobacterial blooms or phytoplankton taxonomic classification. Artificial neural network (ANN) and SVM were used as a classification technique automatic identification of *Scenedesmus coenobia* based on microscope image processing having the precision of 98.63 and 97.32% (Giraldo-Zuluaga et al., 2018). Lately, a Mask-RCNN (Region-based Convolutional Neural Network) model with 85% accuracy has been studied for algae identification in water samples.

2.3.2 AI Mediated Microalgae Screening

The strains are assessed for growth rate, biomass output, the presence of high-value products, and the ability to remove contaminants. For determining biomass, a variety of measurement techniques were used, including straightforward traditional techniques like dry cell weight or manual cell counting (Teng et al., 2020). The deployment of novel strategies in the quickly developing field of biotechnologies depends on the development of sophisticated measurement techniques namely fluorescence measurement-based approaches led to the creation of massive datasets appropriate for AI methodology applications. In a study, single fluorescence emission spectra measurement in conjunction with back-propagation ANN Optimized by Genetic Algorithms was used to explore the viability of deploying ANN for precise monitoring of algal density (GA-BP model) (Liu et al., 2020a, 2020b). The aim of this model was to estimate the cell concentration due to nonlinear relationship between the measured fluorescence emission and the algal cell concentration. The study's findings indicate that using in situ fluorescence spectrometry in conjunction with an ANN to estimate algal cell concentration has a very high potential. Another application of ANN was to distinguish between monoalgal and mixed algal cultures in order to find pollutants in monoalgal cultures. Another application of ANN was to distinguish between monoalgal and mixed algal cultures in order to find pollutants in monoalgal cultures (Franco et al., 2019). The research was based on the measurement of the light absorption spectra of the monoalgal cultures of *Chlorella vulgaris* and *Scenedesmus almeriensis*, and *Chlorella vulgaris*. Table 2.1 summarizes the various isolation techniques explored for microalgae.

2.4 Purification of Microalgae Cultures

2.4.1 Antibiotic Treatment

Antibiotic treatment's primary goal is to reduce bacterial contamination of an algal culture to a level that allows a modest inoculum to transfer enough algal cells without contaminating it with live bacteria. The antibiotic treatment's efficacy and toxicity are dependent on the concentration and duration of exposure, as well as the microalgae and bacterium species (Vu et al., 2018). After 48 hours of exposure to the microalgal culture and antibiotic treatment, the number of viable bacteria drops dramatically. Then, a small percentage of the bacteria can be transferred to a suitable medium without antibiotics. Antibiotics function in one of two ways: bactericidal (e.g., penicillin, vancomycin) or bacteriostatic (e.g., chloramphenicol, tetracycline), which interferes with bacterial metabolism but may not always kill bacteria. Most algae withstand 100/25/25 mg/L of penicillin, streptomycin, and gentamycin, a well-studied and primary antibiotic therapy (Fernandez et al., 2021). Purification can also be accomplished by transferring the algal culture through a succession of antibiotics

Table 2.1 Isolation techniques of microalgae isolated from wastewater

Microalgae	Wastewater	Isolation	Application	References
Chlorosacinopsis sp.	Sewage	Serial dilution	Biofuel	Vasistha et al., (2021a, 2021b)
Scenedesmus sp.	BHU wastewater	Agar streaking	Biofuel	Singh et al. (2022)
Chlorella sorokiniana	Domestic	Agar streaking	Biofuel	Lee et al. (2022)
Chlamydomonas sp.	Rare earth element	Serial dilution, agar streaking	Ammonia nitrogen removal	Zhou et al. (2022)
Chlamydomonas reinhardtii	–	Screening FACS, agar streaking	Recombinant protein	Sproles et al. (2022)
Chlorococcum sp.	Molasses	Serial dilution, agar streaking	Biodiesel	Khanra et al. (2021)
Coalestrella sp.	Sewage	Selective medium with plating	Lipid production	Ferro et al. (2018)
Desmodesmus sp.	Sewage	Selective medium with plating	Lipid production	Ferro et al. (2018)
Chlamydomonas incerta	Agro-industrial wastewater (POME)	Plating method	Biodiesel	Kamyab et al. (2016)
Chlorella sorokiniana	Natural wastewater	TAP plating	Wastewater Treatment	Chiellini et al. (2020)

sequentially, without harming alga. Furthermore, one antibiotic combination may be fatal to some bacteria but suppress the growth of others, but employing antibiotics in sequence may kill the germs that survive (Uddin et al., 2021).

2.4.2 *Enzyme Treatment*

The use of antibiotics to obtain axenic cyanobacteria culture has limitations and occasionally necessitates a different approach. Microalgae are less susceptible to lysozyme digestion than bacteria. A lysozyme basis is another method for creating an axenic culture. Incubating cultures have previously created axenic cyanobacterial cultures for up to 90 minutes at the lowest lethal dose (Vu et al., 2018). Additionally, combining enzymatic therapy with antibiotics can yield promising results. The axenic cultures of the genera *Chlorella* and *Monoraphidium* were obtained using a lysozyme (20 g/mL) and antibiotic cocktail (cefotaxime 500 g/mL and tetracycline 50 g/mL) (Tale et al., 2014).

References

Alam, M.A., Muhammad, G., Rehman, A., Russel, M., Shah, M., & Wang, Z. (2019). Standard techniques and methods for isolating, selecting and monitoring the growth of microalgal strain. *Microalgae Biotechnology for Development of Biofuel and Wastewater Treatment*, 75–93.

Andersen, R. (2005). Recipes for freshwater and seawater media. Algal culturing techniques.

Andersen, R. A., & Kawachi, M. (2005). Microalgae isolation techniques. *Algal Culturing Techniques, 83*, 92.

Andrade, L. M. D., Mendes, M. A., Kowalski, P., & Nascimento, C. A. O. D. (2015). Comparative study of different matrix/solvent systems for the analysis of crude lyophilized microalgal preparations using matrix-assisted laser desorption/ionization time-of-flight mass spectrometry. *Rapid Communications in Mass Spectrometry, 29*(3), 295–303.

Banerjee, C., Bandopadhyay, R., & Shukla, P. (2012). A simple novel agar diffusion method for isolation of indigenous microalgae Chlamydomonas sp. CRP7 and Chlorella sp. CB4 from operational swampy top soil. *Indian Journal of Microbiology, 52*, 710–712.

Barbano, D., Diaz, R., Zhang, L., Sandrin, T., Gerken, H., & Dempster, T. (2015). Rapid characterization of microalgae and microalgae mixtures using matrix-assisted laser desorption ionization time-of-flight mass spectrometry (MALDI-TOF MS). *PLoS ONE, 10*(8), e0135337.

Brennan, L., & Owende, P. (2010). Biofuels from microalgae—A review of technologies for production, processing, and extractions of biofuels and co-products. *Renewable and Sustainable Energy Reviews, 14*(2), 557–577.

Bui, T. V., Ross, I. L., Jakob, G., & Hankamer, B. (2013). Impact of procedural steps and cryopreservation agents in the cryopreservation of chlorophyte microalgae. *PLoS ONE, 8*(11), e78668.

Cadoret, J. P., Bougaran, G., Bérard, J. B., Carrier, G., Charrier, A., Coulombier, N., Garnier, M., Kaas, R., Le Déan, L., Lukomska, E., & Nicolau, E. (2014). Microalgae and biotechnology. *Development of Marine Resources*, 57–115.

Chiellini, C., Guglielminetti, L., Sarrocco, S., & Ciurli, A. (9 Dec 2020). Isolation of four microalgal strains from the lake massaciuccoli: Screening of common pollutants tolerance pattern and perspectives for their use in biotechnological applications. *Front Plant Sci*ence. *11*, 607651. https://doi.org/10.3389/fpls.2020.607651. PMID: 33362836; PMCID: PMC7756032.

Duong, V. T., Li, Y., Nowak, E., & Schenk, P. M. (2012). Microalgae isolation and selection for prospective biodiesel production. *Energies, 5*(6), 1835–1849.

Fernandez-Valenzuela, S., Chávez-Ruvalcaba, F., Beltran-Rocha, J. C., San Claudio, P. M., & Reyna-Martínez, R. (2021). Isolation and culturing axenic microalgae: Mini–review. *The Open Microbiology Journal, 15*(1).

Ferro, L., Gentili, F. G., & Funk, C. (2018). Isolation and characterization of microalgal strains for biomass production and wastewater reclamation in Northern Sweden. *Algal Research, 32*, 44–53.

Fistarol, G. O., Hargreaves, P. I., Walter, J. M., Viana, T. V., Gomes, P. D., Lourenço, C. B., Rezende, C. E., Gregoracci, G., Rua, C., Thompson, C. C., & Thompson, F. L. (2018). Rapid isolation of culturable microalgae from a tropical shallow lake system. *Journal of Applied Phycology, 30*, 1807–1819.

Franco, B. M., Navas, L. M., Gómez, C., Sepúlveda, C., & Acién, F. G. (2019). Monoalgal and mixed algal cultures discrimination by using an artificial neural network. *Algal Research, 38*, 101419.

Giraldo-Zuluaga, J. H., Salazar, A., Diez, G., Gomez, A., Martínez, T., Vargas, J. F., & Peñuela, M. (2018). Automatic identification of Scenedesmus polymorphic microalgae from microscopic images. *Pattern Analysis and Applications, 21*, 601–612.
Jahn, M. T., Schmidt, K., & Mock, T. (2014). A novel cost effective and high-throughput isolation and identification method for marine microalgae. *Plant Methods, 10*(1), 1–10.
Kamyab, H., Din, M. F. M., Ponraj, M., Keyvanfar, A., Rezania, S., Taib, S. M., & Majid, M. Z. A. (2016). Isolation and screening of microalgae from agro-industrial wastewater (POME) for biomass and biodiesel sources. *Desalination and Water Treatment, 57*(60), 29118–29125. https://doi.org/10.1080/19443994.2016.1139101
Khanra, A., Vasistha, S., & Kumar, S., et al. (2021). Cultivation of microalgae on unhydrolysed waste molasses syrup using mass cultivation strategy for improved biodiesel. 3 *Biotech 11*, 287. https://doi.org/10.1007/s13205-021-02823-7.
Lee, J. C., Joo, J. H., Chun, B. H., Moon, K., Song, S. H., Kim, Y. J., Lee, S. M., & Lee, A. H. (2022). Isolation and screening of indigenous microalgae species for domestic and livestock wastewater treatment, biodiesel production, and carbon sequestration. *Journal of Environmental Management, 318*, 115648.
Lee, S. Y., Khoiroh, I., Vo, D.-V.N., Kumar, P. S., & Show, P. L. (2021a). Techniques of lipid extraction from microalgae for biofuel production: a review. *Environmental Chemistry Letters, 19*(1), 231–251.
Lee, T. C. H., Chan, P. L., Tam, N. F. Y., Xu, S. J. L., & Lee, F. W. F. (2021b). Establish axenic cultures of armored and unarmored marine dinoflagellate species using density separation, antibacterial treatments and stepwise dilution selection. *Scientific Reports, 11*(1), 1–13.
Liu, J. Y., Zeng, L. H., Ren, Z. H., Du, T. M., & Liu, X. (2020a). Rapid in situ measurements of algal cell concentrations using an artificial neural network and single-excitation fluorescence spectrometry. *Algal Research, 45*, 101739.
Liu, P., Wang, T., Yang, Z., Hong, Y., Xie, X., & Hou, Y. (2020b). Effects of Fe3O4 nanoparticle fabrication and surface modification on Chlorella sp. harvesting efficiency. *Science of the Total Environment, 704*, 135286.
Mutanda, T., Ramesh, D., Karthikeyan, S., Kumari, S., Anandraj, A., & Bux, F. (2011). Bioprospecting for hyper-lipid producing microalgal strains for sustainable biofuel production. *Bioresource Technology, 102*(1), 57–70.
Pereira, H., Barreira, L., & Custódio, L., et al. (2013). Isolation and fatty acid profile of selected microalgae strains from the Red Sea for biofuel production. *Energies, 6*(6), 2773–2783.
Price, C. A. (2014). *Centrifugation in density gradients*. Academic Press.
Singh, P., Gupta, S.K., Guldhe, A., Rawat, I., & Bux, F. (2015). Microalgae isolation and basic culturing techniques. In *Handbook of marine microalgae* (pp. 43–54). Academic Press.
Singh, R. P., Yadav, P., Kumar, A., Hashem, A., Al-Arjani, A. B. F., Abd-Allah, E. F., Dorantes, A. R., & Gupta, R. K. (2022). Physiological and biochemical responses of bicarbonate supplementation on biomass and lipid content of green algae Scenedesmus sp. BHU1 isolated from wastewater for renewable biofuel feedstock. *Frontiers in Microbiology, 13.*
Sproles, A. E., Berndt, A., Fields, F. J., & Mayfield, S. P. (2022). Improved high-throughput screening technique to rapidly isolate Chlamydomonas transformants expressing recombinant proteins. *Applied Microbiology and Biotechnology, 106*(4), 1677–1689.
Tale, M., Ghosh, S., Kapadnis, B., & Kale, S. (2014). Isolation and characterization of microalgae for biodiesel production from Nisargruna biogas plant effluent. *Bioresource Technology, 169*, 328–335.
Teng, S. Y., Yew, G. Y., Sukačová, K., Show, P. L., Máša, V., & Chang, J. S. (2020). Microalgae with artificial intelligence: A digitalized perspective on genetics, systems and products. *Biotechnology Advances, 44*, 107631.
Thao, T. Y., Linh, D. T. N., Si, V. C., Carter, T. W., & Hill, R. T. (2017). Isolation and selection of microalgal strains from natural water sources in Viet Nam with potential for edible oil production. *Marine Drugs, 15*(7), 194.

Uddin, T.M., Chakraborty, A.J., Khusro, A., Zidan, B.R.M., Mitra, S., Emran, T.B., Dhama, K., Ripon, M.K.H., Gajdács, M., Sahibzada, M.U.K. and Hossain, M.J. (2021). Antibiotic resistance in microbes: History, mechanisms, therapeutic strategies and future prospects. *Journal of infection and public health*, *14*(12), 1750–1766.

Vasistha, S., Khanra, A., Clifford, M., & Rai, M. P. (2021a). Current advances in microalgae harvesting and lipid extraction processes for improved biodiesel production: A review. *Renewable and Sustainable Energy Reviews, 137*, 110498.

Vasistha, S., Khanra, A., & Rai, M. P. (2021b). Influence of microalgae-ZnO nanoparticle association on sewage wastewater towards efficient nutrient removal and improved biodiesel application: An integrated approach. *Journal of Water Process Engineering, 39*, 101711.

Vu, C. H. T., Lee, H. G., Chang, Y. K., & Oh, H. M. (2018). Axenic cultures for microalgal biotechnology: Establishment, assessment, maintenance, and applications. *Biotechnology Advances, 36*(2), 380–396.

Chapter 3
Classification of Microalgae and Factors Responsible for Growth

A few strains have been commercially used so far for the creation of fine chemicals, nutraceuticals, food additives, and aquaculture feed. The classification of algae has eluded applied phycologists. In this chapter, we aim to throw a light on classifying microalgae in a hierarchical, phylogenetically relevant manner, and we attempt to solve this conundrum in this chapter.

3.1 Classification of Microalgae

On the basis of cellular structure and pigments, morphology algae can be classified and presented in Fig. 3.1 (Wehr, 2007).

3.1.1 Chlorophyta (Green Algae)

In the fossil history, chlorophytes emerged billion years ago and are the deepest occurring; erect alga was attached to bedrock at 157 meters depth in the Bahamas. Mostly, chlorophytes are marine, but few can grow on snow, soils in the symbiotic association with protozoans and varying in size from microscopic to large up to several meters (Hollar, 2011). Due to the majority of chlorophyll, most chlorophytes are seen in green color; some may look yellow-green or blackish green owing to the presence of carotenoids or elevated chlorophyll concentration. Furthermore, the cell wall of chlorophytes composed of cellulose, polysaccharides, and proteins accumulates starch as food reservoirs inside plastids (Garcia-Pichel & Belnap, 2021).

M. P. Rai and S. Vasistha, *Microalgae Biotechnology for Wastewater Treatment, Resource Recovery and Biofuels*, https://doi.org/10.1007/978-3-031-31674-6_3

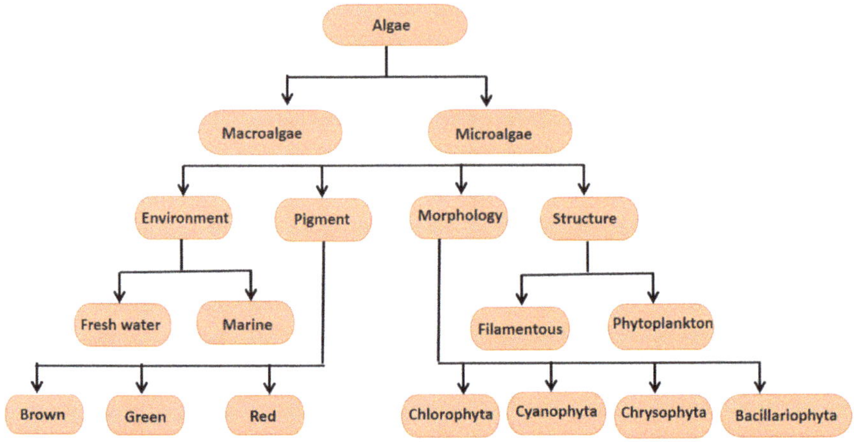

Fig. 3.1 General classification of algae

3.1.2 Cyanophyta (Blue-Green Algae)

Photosynthetic blue-green algae are known as cyanobacteria and classified as bacteria due to lack of membrane-bound nucleus. They tend to elevate the oxygen level in the atmosphere (Seckbach, 2007) and can change their appearance. Various cyanophytes are found in soil and fresh water, and some occur in symbiotic association with protozoa, fungi, and diatoms (Barsanti and Gualtieri, 2022). Phycobilins pigment is produced by blue-green algae and bound to the granules of protein which are connected with the photosynthetic membrane. In addition, they fix the nitrogen and grow with water, sunlight, and nutrients; some freshwater cyanophytes kill the aquatic organism through toxins.

3.1.3 Chrysophyta (Golden Algae)

They are defined as unicellular and photosynthetic organisms containing pigments, including chlorophyll a, c, and fucoxanthin, found in marine and freshwater. Morphologically, brown and golden algae are similar, and both store food in the form of polysaccharide laminarin. In addition, the toxin produced by them is responsible for harming livestock.

3.1.4 *Bacillariophyta (Diatoms)*

Unicellular organism diatoms are majorly planktonic, while few of them develop on other algae. Controlled secretions assist the movement of diatoms by physical and chemical stimuli (Seckbach, 2007). Furthermore, unique shells enclose the protoplasm of diatoms, are made of polymerized silica, and serve as cell walls of diatoms (Hollar, 2011). Bacillariophytes contain pigment chlorophyll a, c, and fucoxanthin and store lipids triacylglycerol (TAG) and a β-1,3 linked carbohydrate termed chrysolaminarin. The mode of reproduction in diatoms is asexual, and they do not have flagella, excluding male gametes.

3.2 Factors Responsible for Microalgal Growth

Numerous factors like light, temperature, pH, and nutrients influence the overall yield of the microalgae biomass and co-products.

3.2.1 *Light*

Light intensity is the main factor for microalgal growth; the duration and intensity are critically responsible for influencing the biochemical composition and biomass rate. The higher and the lower light intensity are not beneficial for microalgae production. When the CO_2 uptake rate is similar to microalgae's respiratory CO_2 release rate, the net biomass growth will be zero. Light is essential for synthesizing ATP and NADPH, which initiate the dark reactions in photosynthesis and produce carbon skeletons (Gao et al., 2022). Some studies demonstrate that a 16 h light and 8 h dark photoperiod is highly appropriate for microalgae growth. *Chlorella Vulgaris* reported high biomass production at a light intensity of 5000–7000 lx (Khan et al., 2022). Furthermore, some microalgae undergo the photoinhibition process due to non-uniform and proper light penetration, which hampers the growth of microalgae. Hence, researchers focus on LED lights or fluorescent tubes to overcome this bottleneck. Photoinhibition can be avoided by providing a high light intensity and continuous shaking of the culture. The optimal light intensity required for higher biomass production lies in 200–400 μM photons/m^2/s (Xiaogang et al., 2020). Longer photoperiods resulted in greater growth of microalga with higher cell densities, according to research on the impact of photoperiod on the growth of Dunaliella salina CCAP 19/30 (Xu et al., 2016). In a study, *Chlorella vulgaris's* light was controlled at 3960, 7920, and 11,920 lux without regard to pH. The light/dark cycle lasted for 12 hours. It was discovered that cells grew at their fastest under 3960 lux (Gong et al., 2014). *Dunaliella salina* CCAP 19/30 have the ability to alter their photosystems in order to maximize photosynthesis. Cells showed phototoxicity when the light intensity was raised to approx 74,000 lux (Xu et al., 2016).

3.2.2 pH and Salinity

Another major factor for boosting biomass production is maintaining the pH of the culture medium (Qiu et al., 2017). Generally, a pH range from 6 to 8.7 is suitable for promoting various microalgae growth, although the pH range varies from species to species. The study showed that *Chlorella Vulgaris* could tolerate a wide range of temperatures, as maximum biomass was observed at pH 9–10 (Mehta et al., 2018). Elevated pH will increase the salinity of the medium, which turns out to be harmful to microalgae. With high pH, ions of carbonates and bicarbonate are accessible to the plants in water. The pH of microalgal cytoplasm is neutral or slightly alkaline and enzymes are pH-sensitive, and the pH can have a direct impact on the microalgae (Chowdury et al., 2020). Researchers examined Scenedesmus sp growth in a pH range from 7 to 10 and found that the ideal pH for this species was 8 (Triphati et al., 2015). *Scenedesmus* sp. LX1. growth by adjusting the pH from 5 to 11 and finding that there was no discernible change in the cultures' growth for pH values of 7, 9, and 11. The microalgae's ability to develop, though, was significantly constrained at pH 5 (Wu et al., 2016).

3.2.3 Temperature

Temperature affects the biochemical components in the microalgae system. The temperature may vary from species to species and culture medium. Elevating the temperature to the optimum range may boost the biomass like temperature below 16 °C retard the algal growth, or more than 36 °C can halt the activity (Zhao et al., 2020). However, some exceptions were observed in the literature as *Anacystis nidulans* and *Chaetoceros* thermophilic microalgae can resist the temperature till 40 °C. In general, most microalgae require a temperature between 20 m and 30 °C (Afzaal et al., 2022). Maintaining a lower temperature disturbs the photosynthesis mechanism by compromising the carbon accumulation activity; on the other hand, the elevated temperature is responsible for deactivating the photosynthesis process; hence, cell shrinks, and eventually, the growth rate declines (Bhattacharya, 2022). This significant impact on photosynthesis is owing to the slump of the enzyme activity ribulose-1–5 biphosphate; it acts as oxygenase or carboxylase. Their action depends on the amount of oxygen and CO_2 available in the chloroplast.

3.2.4 Nutrients

Nitrogen, phosphorus, and carbon are essential nutrients for microalgae growth; those concentrations may vary in species of algae. Microalgae utilize nitrates and phosphates in the inorganic form, and augmented quantity of nitrogen supports microalgae growth. Urea is a cost-effective substitute source for additional inorganic nitrogen sources. Carbon such as glycerol and sucrose is also provided to the medium to boost growth. Some trace elements such as Mn, Fe, Zn, Mo, K, and Co also influence the biomass production in microalgae. The literature stated that limiting the nitrogen source affected the biomass as growth was retarded but responsible for the upsurge production of lipids. The availability of nutrients is highly responsible for carbohydrate and lipid synthesis. Moreover, few bacteria are capable of increasing the microalgae productivities. Bacteria degrade the complex nutrients into a more straightforward form that can be readily available for growth, including ammonia. According to study when the nitrogen concentration rose between 0 and 8.8 mmol L^{-1}, the growth rates of the microalgae *Pseudochlorococcum* sp. suddenly surged to about 5–8 gm^2 d1 and subsequently leveled out (Ji et al., 2014).

References

Afzaal, M., Hameed, S., Rasheed, R., & Khan, W.U.D. (2022). Microalgal biofuels: A sustainable pathway for renewable energy. In *Algal Biotechnology* 187–222. Elsevier.

Barsanti, L., & Gualtieri, P. (2022). Algae: Anatomy, Biochemistry, and Biotechnology (3rd ed.). CRC Press. https://doi.org/10.1201/9781003187707.

Bhattacharya, A. (2022). Effect of low temperature stress on photosynthesis and allied traits: A review. *Physiological Processes in Plants Under Low Temperature Stress*, 199–297.

Chowdury, K. H., Nahar, N., & Deb, U. K. (2020). The growth factors involved in microalgae cultivation for biofuel production: A review. *Computational Water, Energy, and Environmental Engineering, 9*(4), 185–215.

Gao, P., Guo, L., Gao, M., Zhao, Y., Jin, C., & She, Z. (2022). Regulation of carbon source metabolism in mixotrophic microalgae cultivation in response to light intensity variation. *Journal of Environmental Management, 302*, 114095.

Garcia-Pichel, F., & Belnap, J. (2021). Cyanobacteria and algae. *Principles and Applications of Soil Microbiology*, 171–189.

Gong, Q., Feng, Y., Kang, L., Luo, M., & Yang, J. (2014). Effects of light and pH on cell density of Chlorella vulgaris. *Energy Procedia, 61*, 2012–2015.

Hollar, S., ed. (2011). A closer look at bacteria, algae, and protozoa. Britannica Educational Publishing.

Ji, B., Zhang, W., Zhang, N., Wang, J., Lutzu, G. A., & Liu, T. (2014). Biofilm cultivation of the oleaginous microalgae Pseudochlorococcum sp. *Bioprocess and Biosystems Engineering, 37*, 1369–1375.

Khan, M. J., Singh, N., Mishra, S., Ahirwar, A., Bast, F., Varjani, S., Schoefs, B., Marchand, J., Rajendran, K., Banu, J. R., & Saratale, G. D. (2022). Impact of light on microalgal photosynthetic microbial fuel cells and removal of pollutants by nanoadsorbent biopolymers: Updates, challenges and innovations. *Chemosphere, 288*, 132589.

Mehta, P., Rani, R., Gupta, R., Mathur, A. S., & Puri, S. K. (2018). Biomass and lipid production of a novel freshwater thermo-tolerant mutant strain of Chlorella pyrenoidosa NCIM 2738 in seawater salinity recycled medium. *Algal Research, 36*, 88–95.

Qiu, R., Gao, S., Lopez, P. A., & Ogden, K. L. (2017). Effects of pH on cell growth, lipid production and CO_2 addition of microalgae Chlorella sorokiniana. *Algal Research, 28*, 192–199.

Seckbach, J., ed. (2007). Algae and cyanobacteria in extreme environments (Vol. 11). Springer Science & Business Media.

Tripathi, R., Singh, J., & Thakur, I. S. (2015). Characterization of microalga Scenedesmus sp. ISTGA1 for potential CO_2 sequestration and biodiesel production. *Renewable Energy, 74*, 774–781.

Wehr, J. D. (2007). Algae: Anatomy, Biochemistry, and Biotechnology by Barsanti, L. & Gualtieri, P. *Journal of Phycology*, *43*(2), 412–414.

Wu, Y. H., Yu, Y., Hu, H. Y., & Zhuang, L. L. (2016). Effects of cultivation conditions on the production of soluble algal products (SAPs) of Scenedesmus sp. LX1. *Algal Research, 16*, 376–382.

Xiaogang, H., Jalalah, M., Jingyuan, W., Zheng, Y., Li, X., & Salama, E. S. (2020). Microalgal growth coupled with wastewater treatment in open and closed systems for advanced biofuel generation. *Biomass Conversion and Biorefinery*, 1–20.

Xu, Y., Ibrahim, I. M., & Harvey, P. J. (2016). The influence of photoperiod and light intensity on the growth and photosynthesis of Dunaliella salina (chlorophyta) CCAP 19/30. *Plant Physiology and Biochemistry, 106*, 305.

Zhao, T., Han, X., & Cao, H. (2020). Effect of temperature on biological macromolecules of three microalgae and application of FT-IR for evaluating microalgal lipid characterization. *ACS Omega, 5*(51), 33262–33268.

Chapter 4
Traditional Cultivation System

4.1 Modes of Microalgae Cultivation

Three major modes of cultivation are discussed in the following section: photoautotrophic, heterotrophic, and mixotrophic (Fig. 4.2). Methods for cultivation of microalgae highly influence the biomass and lipid production.

4.1.1 *Photoautotrophic*

Autotrophic or photoautotrophic is a commonly used energy-saving mode of cultivation, and the process involves the reduction of carbon dioxide to carbohydrate. The metabolic pathway employs the light and inorganic carbon as the energy source which uplifts the microalgae growth (Javed et al., 2019). In autotrophic cultivation, the cells cause the broth turbidity which decreases light penetration or photolimitation inside the culture medium; hence, this drawback results difficulty in attaining the high biomass and lipid production. In autotrophic condition, the lipid content in microalgae lies in the range of 5–68%. Phototrophic cultivation is the most cost-effective method of growing microalgae. The phototrophic culture can be done in both Open ponds and closed bioreactors (Borowitzka et al., 1997). Open pond systems are more advantageous since they are less expensive than photobioreactors, but they can only grow a limited number of microalgae species. Table 4.1 summarizes the microalgae cultivation in different modes.

4.1.2 *Photoautotrophic Open Cultivation System*

The first Open pond cultivation was done in the 1950s. The majority of the Open ponds are of the raceway type, as recommended by Oswald in 1969. The rectangular

M. P. Rai and S. Vasistha, *Microalgae Biotechnology for Wastewater Treatment, Resource Recovery and Biofuels*, https://doi.org/10.1007/978-3-031-31674-6_4

Table 4.1 Major aspects for microalgae cultivation in different modes (Daneshvar et al., 2021; Zhan et al., 2017)

Different aspects	Photoautotrophic	Heterotrophic	Mixotrophic
Energy	Light	No light	Light
Carbon source	Inorganic	Organic	Organic and inorganic
Biomass production	Low	Medium	High
CO_2 Emission	Negative emission sink	Positive emission	Neutral CO_2 produced and consumed
Metabolism inconsistency	No switch between sources	Simultaneously utilization	Switch between sources
Use of renewable source	Yes	Not in principle	Partially
Limiting factor for growth	Light	Oxygen	Light and oxygen
Contamination risk	Low	High	Medium
Harvesting difficulty	High, due to diluted cultures	Low due to dense biomass concentration	Low, as high biomass concentration

channel with the flow from one end to the other is the most frequent design of raceway ponds. The depth and diameter/width of the raceway pond are crucial parameters to consider. Increased pond width may result in a drop in current speed, resulting in less mass movement and mixing. The amount of culture volume employed and light penetration determine the depth and length of the culture. Open culture methods have been shown to be a cost-effective and long-term cultivation method. These need less energy, are simple to maintain and clean, and hence have the potential to deliver a high net energy production (Chisti, 2007). Open pond systems, on the other hand, have several drawbacks, including water loss, inadequate light use, and a huge necessary space. In addition, there is a limited type of algae production, impure culture development, insufficient mixing, and low algal productivity. The technology is limited to biomass productivity (Javed et al., 2019).

4.1.2.1 Types of Open Pond System

Due to their operational viability and straightforward construction, Open pond are relevant in large-scale microalgal culture. Applications determine how Open pond are configured, including their sizes, forms, and types. Raceway, circular-central pivoted, and unstirred ponds are just a few of the structural frameworks that make up Open pond (Klinthong et al., 2015). Unstirred ponds are commercially viable and utilized for fewer types of microalgae, such as *Dunaliella salina*. Unstirred industrial ponds are built in shallow natural water ponds with a depth of less than 1.5 m. The earliest large-scale microalgal cultivation systems are circular-centrally pivoted ponds with revolving agitators, which completely rely on ponds used for wastewater treatment

(Costa et al., 2019). The system architecture of these ponds, which has a maximum restriction of 10,000 m^2, prevents spinning arms from mixing water uniformly in huge ponds. By instantly dissolving the ingredient pouring into the water tank and creating a uniform concentration in the tank, circular tanks come close to perfect mixing. The raceway system, which has a closed loop and oval recirculation channels (0.2–0.5 m deep) with paddlewheel stirrers, is the most often used synthetic system for outdoor industrial microalgal production and wastewater treatment (Brennan & Owende, 2010). The stirrer makes sure that the culture is completely homogenized to support microalgal cultivation. The raceway method was first used to mass cultivate several microalgal strains for the commercial production of *Spirulina*. Due to its low power requirements and ease of maintenance, the raceway system is the most economically viable method for large-scale microalgal production (Masojídek et al., 2011).

4.1.2.2 Factors Affecting an Open System Effectiveness

Open Pond Configuration

Numerous studies have emphasized on reducing the limitations of Open pond system. The biomass output and nutrient uptake efficiency of Open pond system are directly influenced by their structural design (Acién et al., 2013). To allow for light penetration in the liquid media for microalgal growth, Open pond system has modest pond depths ranging from few millimeters to two centimeters (thin-layer ponds) and 15–30 cm (raceways). In the pilot-scale setup, the specified power consumption was reduced from 6 to 4 W g^{-1} (Apel et al., 2015). When compared to the traditional raceway setup, *Scenedesmus* sp. grown in a 1-m-deep top-lit open bioreactor system produced 0.06 g dw L^{-1} day^{-1} of biomass and a 27% higher amount of lipids (Xiaogang et al., 2020). A different study on the cultivation of *Chlorella pyrenoidosa* in an airlift-driven sloping raceway with 1.55° sloping channels revealed 10.52 g m^{-2} day^{-1} areal biomass productivity, which is 37.34% more productive (Huang et al., 2016).

Agitation

Another crucial element for improving mass transfer in open reactors is mixing. It prevents cell aggregation, sedimentation, and gradients in temperature and nutrients. For optimum cell growth, mixing increases light penetration all the way down to the reactor's lowermost depth. Aeration, pumping, or paddle wheels all encourage mixing in Open pond (Zhao & Su, 2014). Shear pressures and mixing in a stirring tank affect the growth and photosynthetic activity (PA) of *Chlamydomonas reinhardtii*, *Chlorella vulgaris*, and *Scenedesmus obliquus*. *Scenedesmus obliquus* and *C. vulgaris*, respectively, displayed 4.8 and 4.0% PA and 71 and 48% greater growth at 126 cm s^{-1} tip speed (Leupold et al., 2012). In order to ensure correct mixing in the conventional raceway ponds used for *C. pyrenoidosa* culture, a simulation model called computational fluid dynamics was used. This model revealed 14.53

g of biomass output per square meter per day (Huang et al., 2015). The culture of freshwater *Chlorella* sp. required the installation of wing baffles for optimum mixing, which led to a 30.11% higher biomass production than the control system (without wing baffles) (Zhang et al., 2015).

CO_2 Concentration

In particular, during periods of stronger irradiation, the amount of CO_2 taken in from the environment is insufficient to satisfy the carbon requirement for microalgal growth in Open pond. A carbon source should be added to the culture to help with the carbon problem. The supply of CO_2 in Open pond is dependent on a number of variables, including the ideal pH, mass transfer coefficient, mixing, sparger type, and gas-liquid contact time (Mendoza et al., 2013). Due to the reduced transmission of oxygen produced by photosynthetic processes into the ecosystem, dissolved oxygen builds up in Open pond. Numerous microalgal species may also experience photosynthetic inhibition as a result of this buildup, which could lead to photooxidation under intense light. High CO_2 levels boost microalgal photosynthetic efficiency and speed up reproduction. At 5.5–6.0 pH, 40% (v/v) CO_2, and 30 °C, *Chlorella* sp. showed high growth. However, microalgae development can be hindered by CO_2 concentrations greater than 5% (v/v) (Cheah et al., 2015). A 35 cm deep, 110 cm long, and 35 cm wide raceway pond's biomass production rose by 29% with the installation of up-down chute baffles (Cheng et al., 2016). In an enhanced aeration and stirring system, *Chlorella* sp. demonstrated 58, 27, 20, and 16% CO_2 bioconversion efficiency at 2, 5, 10, and 15% (v/v) CO_2 (Zhao & Su, 2014).

pH

An important aspect of determining biomass productivity in Open pond is the impact of pH on microalgal culture. Different media have pH values that range from 6 to 8.76, which is the range that most microalgal species prefer. Maintaining optimal growth is possible at an intracellular pH of 7.5 (Suparmaniam et al., 2019). Variations in biological mechanisms cause cell cultures to fail at high pH levels. By including alkaline chemicals in the microalgae, the aeration system should be constructed with pH control (Mata et al., 2010).

Light Intensity

An idealized Open pond design should be able to absorb light that is essential for microalgal culture and ensure that the microalgal cells consistently receive light for effective biomass synthesis. The amount of light available in Open pond is likewise influenced by pond depth and cell density (Razzak et al., 2013). High light levels may result in photoinhibition and reduce the growth of microalgae, and 10–30 mmol

$m^2\ s^{-1}$ of light are optimal for the growth of microalgae (Zeng et al., 2011). By increasing the light intensity up to 400 mmol $m^2\ s^{-1}$, the rate of photosynthetic metabolic activities in microalgal cells was accelerated.

Scenedesmus and *Chlorella* sp. showed high metabolic rate and biomass productivity at 200 mmol $m^2\ s^{-1}$ (Fernandes et al., 2014). High light output makes it possible for light to penetrate very viscous cultures and boost photosynthesis. Less light absorption by the light-harvesting complexes in dense cultures reduces photoinhibition and guards against cell damage. Another study on *Neochloris oleabundans* revealed that a sequential adjustment in light intensity led to a doubling in biomass production (Pires et al., 2012).

The use of light guides in high-rate ponds led to a 3.9-fold improvement in overall output. High light distribution, increased photon conversion efficiency, decreased dark zones, and reduced respiration losses were the primary outcomes after placing light guides in Open ponds (Sivakaminathan et al., 2020).

Temperature

Another significant element influencing microalgal proliferation in Open pond is temperature. Temperatures between 25 and 35 °C are ideal for growing microalgae; anything higher may result in cell damage (Chisti et al., 2013). Marine microalgal species have decreased growth above 28 °C and are more vulnerable to injury at high temperatures. In Open ponds, the temperature is not controlled and depends on evaporation, the amount of sunlight, and the ambient air quality. Depending on the location of the reactor and the time of year, culture temperatures may change. Low temperatures can reduce the rate of photosynthesis because they are unfavorable for the activity of the enzyme ribulose bisphosphate carboxylase oxygenase (RuBisCO). In contrast, high temperatures reduce CO_2 solubility and metabolic rate (Rahaman et al., 2011). RuBisCo enzyme binds to O_2 and induces photorespiration at low CO_2 solubility, which lowers the rate of carbon bioconversion by 20–30%. Regulation of temperature variations and the choice of species that are highly tolerant of temperature changes are so crucial (Slegers et al., 2011).

4.1.3 Photoautotrophic Closed Cultivation System

Photobioreactors are generally closed systems used to grow phototrophic microalgae in which energy is supplied by an artificial light source (Andersen & Kawachi, 2005). They provide uniform dispersion and efficient light use, leading to high gas mass transfer (like CO_2 and O_2). Closed systems typically consist of four phases: solid, liquid, gas, and light (Molina et al., 2001). Microalgae are in the solid phase, growth media are in the liquid phase, CO_2 and O_2 are in the gaseous phase, and light are in the visible phase.

Tubular, flat plate, and column photobioreactors are examples of closed systems (Andersen et al., 2005; Sierra et al., 2008; Slegers et al., 2011). A closed system with a high transparent surface and a low number of lit parts would be optimal. Closed systems are thought to be the most effective for cultivating a specific species in a controlled environment. As closed systems, flat-plate reactors were used. They have a large surface area and microalgae cell densities of more than 80 g/L (Hu et al., 1998). They are made of a transparent material with a high solar energy capture rate, and they have higher photosynthesis efficiency than tubular bioreactors. It is difficult to scale up a tubular reactor. The only way to make a large tubular reactor is to combine smaller units together, which creates operational and maintenance issues. However, major tubular reactors are in use around the world, including a 25 m^3 reactor at Mera Pharmaceuticals in Hawaii and a 700 m^3 reactor at Klötze in Germany (Olaizola, 2000; Pulz et al., 2001). Column bioreactors were presented as a solution to these issues. These offer a high volumetric mass transfer, adjustable growing conditions, a small footprint, and are simple to use.

4.1.4 *Membrane Photobioreactor*

CO_2 delivery has a significant impact on the biomass productivity of microalgae cultivation systems. In most cases, CO_2 is delivered by simple air bubbling (Fig. 4.1). Inadequate mixing results in poor mass transfer and efficiency (Chnag et al., 2016). This issue can be solved by adopting membrane-assisted bioreactors, which have a higher CO_2 mass transfer rate than conventional bioreactors. This technique is carried out in two steps:

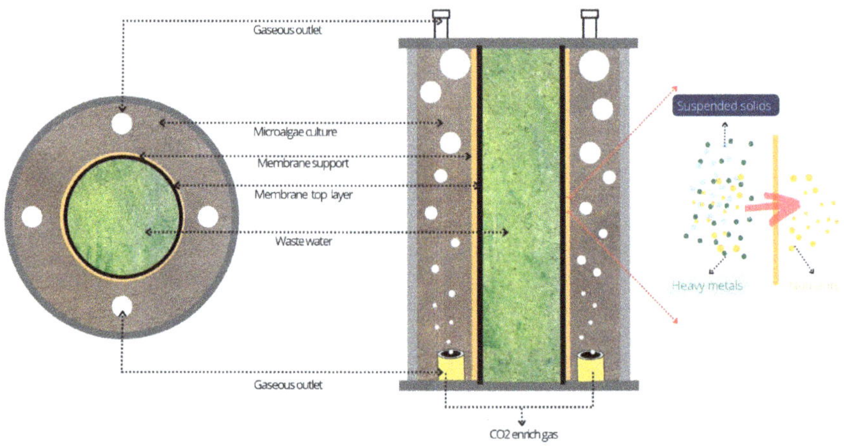

Fig. 4.1 Microalgae cultivation and wastewater treatment in membrane photobioreactor

(1) A membrane is used to continually remove water while biomass is kept on the membrane.
(2) A membrane is utilized to deliver CO_2 either as contacted or sparger throughout cultivation.

Many membrane systems have been investigated for microalgae cultivation over time (Kumar et al., 2010).

Hollow fiber membrane photobioreactors are adopted for microalgae cultivation due to their increased interfacial area and high algal biomass for biofuel conversion from wastewater with high nutrient strength. *Chlorella vulgaris* had the highest biomass productivity of 4 g/m^3/h (Kumar et al., 2010). According to the findings, combining wastewater treatment, CO_2 sequestration, and biofuel generation is a potential path for microalgae development. Different photobioreactors and their biomass productivity are compared.

4.1.4.1 Factors Affecting the Efficiency of Photobioreactor

Reactor Configuration

The photobioreactor arrangement plays a significant role in determining the potential for pollution removal and biomass productivity of microalgal species.

There are some reactor types and designs that are better than others, such as the bubble column reactor, which has good mass transfer and a high surface area to volume ratio; the horizontal tube, which has good CO_2 utilization, temperature control, and mass transfer; the stirred tank reactors, which have adequate mixing and lighting; and the flat panel, which has good illumination surface area and biomass productivity (Chew et al., 2018). For high biomass yield in photobioreactor, tubes with a diameter of 0.1 m are typically used. For the cultivation of the *Chlorella* strain, a study reported improvements to the photobioreactor design (an arrayed cylindrical photobioreactor) and 1.5 g L^{-1} biomass productivity (Huang et al., 2019). The biomass productivity of *Chlorella* sp. was boosted (56%, 0.340 g L^{-1} day^{-1}) in a different investigation on the retrofitting of open-tank photobioreactor with transparent rectangular acrylic chambers (Hsieh et al., 2009a, 2009b). Other variations, including luminescence photobioreactor, baffled integrated photobioreactor, and floating horizontal photobioreactor, have been reported in the literature and are discussed in the subsections with their associated impacts. These changes improve the way light is used, how nutrients are stirred up, and how temperature is managed during microalgal growth. A significant factor influencing biomass productivity is the agitation system, which can be divided into mechanical and non-mechanical types (raceway ponds and stirred tanks in Open pond and bubble column, tubular Photobioreactor, and airlift in Photobioreactor). The operational stability of photobioreactor also relies on the prevention of foreign invasive microalgal species (Kumar et al., 2011). One of the most important features is the cleanability of Photobioreactor since it

(1) prevents the growth of biofilm on the reactor walls, which obstructs light transmission,
(2) lowers the danger of contamination.

Implementing the following principles will boost the cleanability of a Photobioreactor: (1) smooth internal surface; (2) fewer bends and internals; and (3) enormous dimensions (internal) of a Photobioreactor for simple cleaning (Xiaogang et al., 2020).

Light Intensity

Photobioreactor effectiveness is impacted by a variety of lighting parameters, including light absorbance, transport, distribution, and use by photosynthetic microalgae (Greenwell et al., 2010). The three categories of light saturation, light limitation, and light inhibition should not apply to the cellular growth of phototrophic microbes. With an increase in light intensity to 420 mmol m^2 s^{-1}, *S. obliquus* CNW-N produced biomass three times as much (Ho et al., 2012). *Nannochloris atomus* grown in a floating horizontal Photobioreactor at 31.3 k lux light intensity displayed 4.0 g L^{-1} of biomass concentration and 12.9 g m^{-2} day^{-1} biomass productivity (Dogaris et al., 2015). By adding hollow light guides to a flat-plate Photobioreactor, the normal light intensity (70 mol m^2 s^{-1}) was boosted to 6.5 times, allowing researchers to examinc thc impact of light intensity. The findings revealed that *C. vulgaris* FACHB-31's biomass productivity has increased by 23.42% (Sun et al., 2016). The ability to adjust to changes in light intensity exists in microalgal cells. They make a sufficient amount of photosynthetic reaction center antenna complex during light changes, which also protects the cells from photodamage under high light intensity.

CO_2 Concentration and pH

CO_2 concentration and pH also have a significant role in microalgal growth, in addition to agitation and light intensity. CO_2 has the following effects on *Gloeothece membranacea* growth investigated in a photobioreactor for luminescence. According to the findings, 36.6% of lipid productivity was seen with a 15% CO_2 supply (Mohsenpour et al., 2016). When coupled phosphate and bicarbonate buffers were installed to manage the pH and CO_2 concentration in bubble column Photobioreactor as part of an economic investigation of *Haematococcus pluvialis* culture at an industrial scale, a 105% improvement in biomass productivity was noted (Choi et al., 2017). The pH value, which has an ideal range of 7–9, is another significant factor regulating the growth of microalgae.

Some species, however, can thrive in pH ranges that are either acidic or basic. In order to develop the community of microalgal cells in a culture medium, pH is important. The physiochemical characteristics of microalgal cells, such as their metabolic makeup, polarity, and electron transport system, may vary as the pH level fluctuates. *Scenedesmus obtusiusculus* AT-UAM demonstrated 85 mg L^{-1} day^{-1} biomass productivity with 60% fat content when grown in nitrogen deficiency conditions at a regulated pH of 5.5. *Dunaliella salina* was mass-cultivated at pH 9 and produced 17.85 mg L per day of biomass (Khadim et al., 2018).

Temperature

In microalgal development, temperature determines the photosynthetic rate and is crucial for CO_2 fixation. During microalgal cultivation, it controls the activity of enzymes, cell division, and the rate of nutrient absorption. It has been suggested that temperatures between 15 and 30 °C are ideal for microalgal growth. Although some species can thrive in both high and low temperatures (35 °C and 15 °C, respectively), below and above these ranges, microalgal cells either die or grow slowly (Xu et al., 2019). In order to reduce the amount of micropollutants in the wastewater and boost biomass production, *Acutodesmus obliquus* was grown at 30 °C in it (Reymann et al., 2020). The temperature range of 25–30 °C has been substantiated by lipid synthesis from microalgal biomass. The finest *Chaetoceros* sp. FIKU035 cultivar for making biodiesel was thought to be cultivated at 30 °C (Chaisutyakorn et al., 2018). Depending on the type of species, different species respond to temperature changes differently, but for all species, the impact on the fatty acid profile is consistent. Along with the effectiveness of CO_2 sequestration and biomass productivity, the unsaturated fatty acids decrease as the temperature drops (Xu et al., 2019).

4.1.5 Photoautotrophic Hybrid Cultivation System

Microalgae growth in photobioreactors and Open ponds are combined in hybrid farming. Photobioreactors are employed in the initial step, with regulated conditions to reduce the risk of contamination and promote cell division (Rodolfi et al., 2009). Cells are exposed to nutrients in the second stage, which aids in increasing lipid output. The second step should ideally be an Open pond, where environmental factors aid microalgae production. Using *Haematococcus pluvialis*, researchers developed a hybrid system for oil and astaxanthin production, achieving an oil production rate of 10 tons per hectare (Huntley et al., 2007). They also showed that high oil content species may produce 76 tons of oil per hectare each year. The concept of two-stage oil production was proposed, in which one portion of the plant is dedicated to biomass production while nitrogen is abundant and the rest is committed to oil production when nitrogen is scarce. The first section of the plant generated 10 kilogram of lipid/ha/day, and the second section produced 80 kg of lipid/ha/day, for a total production rate of 90 kg of lipid/ha/day. Where average solar radiation is 20 $MJ/M^2/day$, tropical areas of the planet can produce nearly 30 tons of oil each year (Javed et al., 2019).

4.2 Heterotrophic Mode

The metabolic pathway in heterotrophic mode of cultivation involves the utilization of organic compounds as carbon and energy source in absence of light. Various carbon sources in the form of sugar or glucose are most preferred in this type of cultivation, though few cost-effective alternative sources like corn power hydrolysate and glycerol can be used. Heterotrophic cultivation is devoid of photolimitation; therefore, boosted biomass and lipid productivities can be achieved.

The shortcomings associated with heterotrophic mode involve: (i) Absence of light results in less production of metabolities and phytochemicals, (ii) expensive process as the large amount of organic carbon is required, and (iii) the process may reduce the quality of the product, due to contamination by microorganism as they are fast growers (De et al., 2003).

4.3 Mixotrophic Mode

This mode of cultivation utilizes both inorganic and organic source of carbon and results in augmented biomass and lipid production. In mixotrophic process, the specific growth rate of microalgae is the sum of both modes of cultivation, i.e., photoautotrophic and heterotrophic (Zhang et al., 1999).

The mixotrophic mode offers the longer exponential phase which results in higher biomass production. In addition, this process reduces the loss of biomass caused from respiration during dark hours. Pigments such as β-carotene, astaxanthin, phycobilins, fucoxanthin, and zeaxanthin are conserved during mixotrophic mode (Andrade & Costa, 2007). The limitations associated with mixotrophic cultivation include (i) the requirement of organic and inorganic carbon source, sometimes the process become expensive, and (ii) high chance of contamination (Javed et al., 2019). Table 4.1 summarizes microalgae cultivation in different modes (Fig. 4.2; Table 4.2).

4.4 Microalgae Cultivation Strategies

The chosen culture strategy determines the success of the production of microalgae-derived biofuels. The three basic categories of culture techniques are batch culture, semi-continuous culture, and continuous culture (Zhu et al., 2015). All of the techniques employed in microalgal cultivation have advantages and disadvantages in terms of lipid productivity and economic viability, which are discussed in the following sections.

Table 4.2 Summary of microalgae cultivation studies in different modes for biomass production

Microalgae	Mode of cultivation	Source	Biomass production (g/L)	References
Chlorella protothecoides	Heterotrophic	Fructose	3.90	(Gao et al., 2010)
	Heterotrophic	Glucose	3.74	(Xu et al., 2006)
	Heterotrophic	Sucrose	1.20	(Gao et al., 2010)
Chlorella sorokiniana	Mixotrophic	Acetate	1.10	(Ogbonna et al., 2000)
	Heterotrophic	Acetate	0.11	(Wan et al., 2011)
	Mixotrophic	Glucose	1.32	
	Heterotrophic	Glucose	0.09	(Choix et al., 2014)
Chlorella vulgaris	Mixotrophic	Glucose	1.175	
	Mixotrophic	Sucrose	0.525	
	Mixotrophic	Xylose	0.307	(Kong et al., 2020)
	Mixotrophic	Glycerol	0.354	
	Photoautotrophic	CO_2	0.367	
	Heterotrophic	Glucose	0.956	
	Heterotrophic	Sucrose	0.452	
	Heterotrophic	Acetate	0.24	(Perez-Garcia et al., 2011)
	Heterotrophic	Urea	0.10	(Perez-Garcia et al., 2010)
	Mixotrophic	Glucose	1.31	(Nouri et al., 2021)
	Mixotrophic	Glucose	1.7	(Zhang et al., 2020)
	Heterotrophic	Glucose	3.00	(Cai et al., 2021)
Dunaliella salina	Mixotrophic	Glucose	0.52	(Wan et al., 2011)
NannochlorOpen pond is Sp.	Photoautotrophic	CO_2	0.40	(Wan et al., 2011)
	Mixotrophic	Glucose	3.022	(Patel et al., 2019)
Chlorella protothecoides	Mixotrophic	Acetate	2.14	
	Mixotrophic	Glycerol	0.7	
Scenedesmus sp.	Mixotrophic	Acetate	0.80	(Park et al., 2012)

(continued)

Table 4.2 (continued)

Microalgae	Mode of cultivation	Source	Biomass production (g/L)	References
Scenedesmus obliquus	Photoautotrophic	CO_2	0.080	(Park et al., 2012)
Thraustochytrium	Heterotrophic	Glycerol	4.04	(Chen et al., 2020)
	Mixotrophic	Acetate	1.81	(Chen, 1996)
Spirulina platensis	Mixotrophic	Molasses	2.90	(Andrade & Costa, 2007)

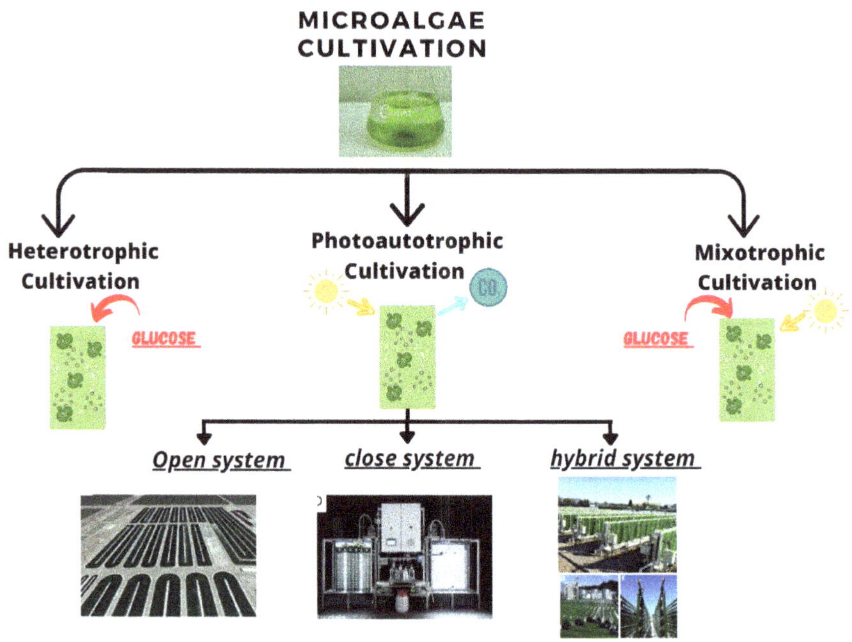

Fig. 4.2 Microalgae cultivation modes

4.4.1 Batch Cultivation

In this cultivation, the microalgae are grown for a pre-determined amount of time under precise environmental conditions in a given volume of nutrient culture. The biomass of *Micractinium inermum* NLP-F014 rose everyday under batch cultivation until day 10, when it reached 5.10 g L^{-1}, at which point the growth slightly slowed down as a result of nutrient consumption (Eldiehy et al., 2022). In this method, a variety of variables, including nutrient concentrations, pH, temperature, and dissolved oxygen, directly affect microalgae growth. Of all the growing techniques, the batch approach is the most straightforward, adaptable, and dependable (Hsieh et al., 2009a, 2009b). The main drawbacks of this approach, on the other hand, are the challenges of maintaining strains in the continuous exponential phase, the rapid nutrient consumption, the restriction to a single cultivation cycle, and the uneven exposure to irradiance due to the cell self-shading effect, which ultimately results in decreased biomass yield (Alazaiza et al., 2023). Researchers recommended alternate methods, including continuous and semi-continuous cultivation, to alleviate these drawbacks.

4.4.2 Continuous Cultivation

The primary characteristic of this kind is that algal development occurs under steady-state circumstances, meaning that cell division happens steadily in a controlled environment (Tang et al., 2012). Additionally, the researcher has complete autonomy over all process parameters.

The continuous culture approach initially resembles batch cultivation in that both are given the nutrients needed at the start of cultivation. However, in continuous cultivation, new media are eventually introduced to maintain the algae in a log phase, allowing cells to expand at an unpredictable rate. As a result, the biomass of algae increases dramatically (Suparmaniam et al., 2019). By employing 10.0%, 5.0%, 7.50%, and 7.50% of landfill leachate, respectively, as a food source during continuous cultivation, the highest amounts of carbohydrates, lipids, proteins, and biomass were obtained from *Chlorella minutissima* 26a (Eldiehy et al., 2022).

The key benefits of this method include higher levels of automation, large biomass densities, and easy management of nutrients and physical elements to deliver a steady supply of high-quality cells. Apart from that, this method has some limitations that prevent its application, such as complexity, difficulties in maintenance, and high equipment costs (Zhu et al., 2015).

4.4.3 Semi-continuous Cultivation

By using the remaining algal cultures as the starting inoculum for a new batch and supplementing them with all the nutrients needed to establish the new culture, this method routinely discards some mature algal culture (Zhu et al., 2015). In addition to helping strains stay in the exponential phase for a longer amount of time than batch culture, this method can create a large number of subsequent cycles. The overall amount of biomass produced rises as a result of this. The piggery wastewater was used as a culture medium, and microalgal consortium was cultivated for 112 days (six cycles of 13 days each) in semi-continuous culture mode led to the highest levels of cellular biomass, carbohydrates, lipids, and proteins of 4.98×106 cell/mL, 157.80,

189.78, and 610.28 mg L^{-1}, respectively (Fernández-Linares et al., 2020). However, in other instances, it is impossible to carry on the cultivations because of the rising levels of contaminants, competitors, and metabolic waste.

Monoraphidium dybowskii Y2 and *Chlorella* sp. L1 were grown in a 40,000 L open raceway pond using batch and semi-continuous cultivation methods (He et al., 2016). The results showed that *M. dybowskii* and *Chlorella* sp. had the maximum biomass productivity in the semi-continuous mode, at 18.23 and 19.35 g m 2 d^{-1}, respectively, as opposed to 9.94 and 17.03 g m^{2} d^{-1} in the batch mode. Additionally, in *M. dybowskii* and *Chlorella* sp., the expenses for lipid production in the semi-continuous mode were 13.31 $ gal and 14.18 $ gal, respectively, as opposed to 21.0 $ gal and 16.54 $ gal in the batch method. In comparison with batch culture, semi-continuous and continuous cultivation often yields a faster growth rate with less wastewater formation (water is used longer) (Ruiz-Marin et al., 2010). The high energy need and unsure stability for long-term operating performances are just two of the problems with these techniques that need to be fixed. Therefore, for commercial microalgae applications, the development of enhanced growing methodologies is required (Zhu et al., 2015).

References

Acién Fernández, F. G., Fernández Sevilla J. M., Molina Grima E. (2013). Photobioreactors for the production of microalgae. Rev Environ Sci Biotechnol, *12*(2), 131–151.

Alazaiza, M. Y., Albahnasawi, A., Al Maskari, T., Abujazar, M. S. S., Bashir, M. J., Nassani, D. E., & Abu Amr, S. S. (2023). Biofuel production using cultivated algae: Technologies, economics, and its environmental impacts. *Energies, 16*(3), 1316.

Andersen, R. A., & Kawachi, M. (2005). Microalgae isolation techniques. *Algal culturing techniques, 83*, 92.

Andrade, M. R., & Costa, J. A. V. (2007). Mixotrophic cultivation of microalga *Spirulina platensis* using molasses as organic substrate. *Aquaculture, 264*(1), 130–134.

Apel, A. C., & Weuster-Botz, D. (2015). Engineering solutions for open microalgae mass cultivation and realistic indoor simulation of outdoor environments. *Bioprocess and Biosystems Engineering, 38*, 995–1008.

Borowitzka, M. A. (1997). Microalgae for aquaculture: opportunities and constraints. *Journal of Applied Phycology, 9*, 393–401.

Brennan, L., & Owende, P. (2010). Biofuels from microalgae—A review of technologies for production, processing, and extractions of biofuels and co-products. *Renewable and Sustainable Energy Reviews, 14*(2), 557–577.

Cai, Y., Liu, Y., Liu, T., Gao, K., Zhang, Q., Cao, L., Wang, Y., Wu, X., Zheng, H., Peng, H., & Ruan, R. (2021). Heterotrophic cultivation of *Chlorella vulgaris* using broken rice hydrolysate as carbon source for biomass and pigment production. *Bioresource Technology, 323*, 124607.

Chaisutyakorn, P., Praiboon, J., & Kaewsuralikhit, C. (2018). The effect of temperature on growth and lipid and fatty acid composition on marine microalgae used for biodiesel production. *Journal of Applied Phycology, 30*(1), 37–45.

Cheah, W. Y., Show, P. L., Chang, J.-S., Ling, T. C., & Juan, J. C. (2015). Biosequestration of atmospheric CO_2 and flue gas-containing CO_2 by microalgae. *Bioresource Technology, 184*, 190–201.

Chen, C.-Y., Lee, M.-H., Dong, C.-D., Leong, Y. K., & Chang, J.-S. (2020). Enhanced production of microalgal lipids using a heterotrophic marine microalga Thraustochytrium sp. BM2. *Biochemical Engineering Journal, 154*, 107429.

Chen, F. (1996). High cell density culture of microalgae in heterotrophic growth. *Trends in Biotechnology, 14*(11), 421–426.

Cheng, J., Yang, Z., Ye, Q., Zhou, J., & Cen, K. (2016). Improving CO_2 fixation with microalgae by bubble breakage in raceway ponds with up-down chute baffles. *Bioresource Technology, 201*, 174–181.

Chew, K. W., Chia, S. R., Show, P. L., Yap, Y. J., Ling, T. C., & Chang, J.-S. (2018). Effects of water culture medium, cultivation systems and growth modes for microalgae cultivation: A review. *Journal of the Taiwan Institute of Chemical Engineers, 91*, 332–344.

Chisti, Y. (2013). Raceways-based production of algal crude oil. *Green, 3*(3–4). https://doi.org/10.1515/green-2013-001861

Chisti, Y. (2007). Biodiesel from microalgae. *Biotechnology Advances, 25*(3), 294–306.

Choi, Y. Y., Joun, J. M., Lee, J., Hong, M. E., Pham, H.-M., Chang, W. S., & Sim, S. J. (2017). Development of large-scale and economic pH control system for outdoor cultivation of microalgae *Haematococcus pluvialis* using industrial flue gas. *Bioresource Technology, 244*, 1235–1244.

Choix, F. J., Bashan, Y., Mendoza, A., & de-Bashan, L. E. (2014). Enhanced activity of ADP glucose pyrophosphorylase and formation of starch induced by *Azospirillum brasilense* in *Chlorella vulgaris*. *Journal of Biotechnology, 177*, 22–34.

Costa, J. A. V., Freitas, B. C. B., Santos, T. D., Mitchell, B. G., & Morais, M. G. (2019). Open pond systems for microalgal culture. In *Biofuels from algae* (pp. 199–223). Elsevier.

Daneshvar, E., Sik Ok, Y., Tavakoli, S., Sarkar, B., Shaheen, S. M., Hong, H., Luo, Y., Rinklebe, J., Song, H., & Bhatnagar, A. (2021). Insights into upstream processing of microalgae: A review. *Bioresource Technology, 329*, 124870.

De Swaaf M. E., Sijtsma, L., & Pronk J. T. (2003). High-cell-density fed-batch cultivation of the docosahexaenoic acid producing marine alga Crypthecodinium cohnii. *Biotechnol Bioeng, 81*(6), 666–672.

Dogaris, I., Welch, M., Meiser, A., Walmsley, L., & Philippidis, G. (2015). A novel horizontal photobioreactor for high-density cultivation of microalgae. *Bioresource Technology, 198*, 316–324.

Eldiehy, K. S., Bardhan, P., Borah, D., Gohain, M., Rather, M. A., Deka, D., & Mandal, M. (2022). A comprehensive review on microalgal biomass production and processing for biodiesel production. *Fuel, 324*, 124773.

Fernandes, B. D., Mota, A., Ferreira, A., Dragone, G., Teixeira, J. A., & Vicente, A. A. (2014). Characterization of split cylinder airlift photobioreactors for efficient microalgae cultivation. *Chemical Engineering Science, 117*, 445–454.

Fernández-Linares, L.C., Gutiérrez-Márquez, A., & Guerrero-Barajas, C. (2020). Semi-continuous culture of a microalgal consortium in open ponds under greenhouse conditions using piggery wastewater effluent. *Bioresource Technology Reports, 12*, 100597.

Gao, C., Zhai, Y., Ding, Y., & Wu, Q. (2010). Application of sweet sorghum for biodiesel production by heterotrophic microalga *Chlorella protothecoides*. *Applied Energy, 87*(3), 756–761.

Greenwell, H. C., Laurens, L. M. L., Shields, R. J., Lovitt, R. W., & Flynn, K. J. (2010). Placing microalgae on the biofuels priority list: A review of the technological challenges. *Journal of the Royal Society, Interface, 7*(46), 703–726.

He, Q., Yang, H., & Hu, C. (2016). Culture modes and financial evaluation of two oleaginous microalgae for biodiesel production in desert area with open raceway pond. *Bioresource Technology, 218*, 571–579.

Ho, S. H., Chen, C. Y., & Chang, J. S. (2012). Effect of light intensity and nitrogen starvation on CO_2 fixation and lipid/carbohydrate production of an indigenous microalga *Scenedesmus obliquus* CNW-N. *Bioresource Technology, 113*, 244–252.

Hsieh, C.-H., & Wu, W.-.T. (2009a). A novel photobioreactor with transparent rectangular chambers for cultivation of microalgae. *Biochemical Engineering Journal, 46*(3), 300–305.

Hsieh, C. H., & Wu, W. T. (2009b). Cultivation of microalgae for oil production with a cultivation strategy of urea limitation. *Bioresource Technology*, *100*(17), 3921–3926.

Huang, J., Hankamer, B., & Yarnold, J. (2019). Design scenarios of outdoor arrayed cylindrical photobioreactors for microalgae cultivation considering solar radiation and temperature. *Algal Research, 41*, 101515.

Huang, J., Qu, X., Wan, M., Ying, J., Li, Y., Zhu, F., Wang, J., Shen, G., Chen, J., & Li, W. (2015). Investigation on the performance of raceway ponds with internal structures by the means of CFD simulations and experiments. *Algal Research, 10*, 64–71.

Huang, J., Yang, Q., Chen, J., Wan, M., Ying, J., Fan, F., Wang, J., Li, W., & Li, Y. (2016). Design and optimization of a novel airlift-driven sloping raceway pond with numerical and practical experiments. *Algal Research, 20*, 164–171.

Huntley, M. E., & Redalje, D. G. (2007). CO_2 mitigation and renewable oil from photosynthetic microbes: a new appraisal. *Mitigation and Adaptation Strategies for Global Change, 12*(4), 573–608.

Hu Q., Kurano, N., Kawachi, M., Iwasaki, I., Miyachi S. (1998). Ultrahigh-cell-density culture of a marine green alga Chlorococcum littorale in a flat-plate photobioreactor. *Appl Microbiol Biotechnol, 49*(6), 655–662.

Javed, F., Aslam, M., Rashid, N., Shamair, Z., Khan, A. L., Yasin, M., Fazal, T., Hafeez, A., Rehman, F., Rehman, M. S. U., & Khan, Z. (2019). Microalgae-based biofuels, resource recovery and wastewater treatment: a pathway towards sustainable biorefinery. *Fuel, 255*, 115826.

Khadim, S. R., Singh, P., Singh, A. K., Tiwari, A., Mohanta, A., & Asthana, R. K. (2018). Mass cultivation of *Dunaliella salina* in a flat plate photobioreactor and its effective harvesting. *Bioresource Technology, 270*, 20–29.

Klinthong, W., Yang, Y. H., Huang, C. H., & Tan, C. S. (2015). A review: microalgae and their applications in CO_2 capture and renewable energy. *Aerosol and Air Quality Research, 15*(2), 712–742.

Kong, W., Yang, S., Wang, H., Huo, H., Guo, B., Liu, N., Zhang, A., & Niu, S. (2020). Regulation of biomass, pigments, and lipid production by *Chlorella vulgaris* 31 through controlling trophic modes and carbon sources. *Journal of Applied Phycology, 32*(3), 1569–1579.

Kumar, A., Yuan, X., Sahu, A. K., Dewulf, J., Ergas, S. J., & Van Langenhove, H. (2010). A hollow fiber membrane photo-bioreactor for CO_2 sequestration from combustion gas coupled with wastewater treatment: A process engineering approach. *Journal of Chemical Technology and Biotechnology, 85*(3), 387–394.

Kumar, K., Dasgupta, C. N., Nayak, B., Lindblad, P., & Das, D. (2011). Development of suitable photobioreactors for CO_2 sequestration addressing global warming using green algae and cyanobacteria. *Bioresource Technology, 102*(8), 4945–4953.

Leupold, M., Hindersin, S., Gust, G., Kerner, M., & Hanelt, D. (2012). Influence of mixing and shear stress on *Chlorella vulgaris*, *Scenedesmus obliquus*, and *Chlamydomonas reinhardtii*. *Journal of Applied Phycology V, 25*(2), 485–495.

Masojídek, J., Kopecký, J., Giannelli, L., & Torzillo, G. (2011). Productivity correlated to photobiochemical performance of Chlorella mass cultures grown outdoors in thin-layer cascades. *Journal of Industrial Microbiology and Biotechnology, 38*(2), 307–317.

Mata, T. M., Martins, A. A., & Caetano, N. S. (2010). Microalgae for biodiesel production and other applications: A review. *Renewable and Sustainable Energy Review, 14*(1), 217–232.

Mendoza, J. L., Granados, M. R., de Godos, I., Acien, F. G., Molina, E., Heaven, S., & Banks, C. J. (2013). Oxygen transfer and evolution in microalgal culture in open raceways. *Bioresource Technology, 137*, 188–195.

Mohsenpour, S. F., & Willoughby, N. (2016). Effect of CO_2 aeration on cultivation of microalgae in luminescent photobioreactors. *Biomass and Bioenergy, 85*, 168–177.

Molina, E., Fernández, J., Acién, F., & Chisti, Y. (2001). Tubular photobioreactor design for algal cultures. *Journal of Biotechnology, 92*(2), 113–131.

Nouri, H., Mohammadi Roushandeh, J., Hallajisani, A., Golzary, A., & Daliry, S. (2021). The effects of glucose, nitrate, and pH on cultivation of Chlorella sp. Microalgae. *Global Journal of Environmental Science and Management, 7*(1), 103–116.

Ogbonna, J. C., Yoshizawa, H., & Tanaka, H. (2000). Treatment of high strength organic wastewater by a mixed culture of photosynthetic microorganisms. *Journal of Applied Phycology, 12*(3), 277–284.

Olaizola, M. (2000). Commercial production of astaxanthin from Haematococcus pluvialis using 25,000-liter outdoor photobioreactors. *Journal of Applied Phycology, 12*(3–5), 499–506.

Park, K. C., Whitney, C., McNichol, J. C., Dickinson, K. E., MacQuarrie, S., Skrupski, B. P., Zou, J., Wilson, K. E., O'Leary, S. J. B., & McGinn, P. J. (2012). Mixotrophic and photoautotrophic cultivation of 14 microalgae isolates from Saskatchewan, Canada: potential applications for wastewater remediation for biofuel production. *Journal of Applied Phycology, 24*(3), 339–348.

Patel, A. K., Joun, J. M., Hong, M. E., & Sim, S. J. (2019). Effect of light conditions on mixotrophic cultivation of green microalgae. *Bioresource Technology, 282*, 245–253.

Perez-Garcia, O., Bashan, Y., & Esther Puente, M. (2011). Organic carbon supplementation of sterilized municipal wastewater is essential for heterotrophic growth and removing ammonium by the microalga *Chlorella vulgaris* 1. *Journal of Phycology, 47*(1), 190–199.

Perez-Garcia, O., De-Bashan, L. E., Hernandez, J.-P., & Bashan, Y. (2010). Efficiency of growth and nutrient uptake from wastewater by heterotrophic, autotrophic, and mixotrophic cultivation of *Chlorella vulgaris* immobilized with *Azospirillum brasilense* 1. *Journal of Phycology, 46*(4), 800–812.

Pires, J. C. M., Alvim-Ferraz, M. C. M., Martins, F. G., & Simões, M. (2012). Carbon dioxide capture from flue gases using microalgae: Engineering aspects and biorefinery concept. *Renewable and Sustainable Energy Review, 16*(5), 3043–3053.

Pulz, O. (2001). Photobioreactors: Production systems for phototrophic microorganisms. *Applied Microbiology and Biotechnology, 57*(3), 287–293.

Rahaman, M. S. A., Cheng, L.-H., Xu, X.-H., Zhang, L., & Chen, H.-L. (2011). A review of carbon dioxide capture and utilization by membrane integrated microalgal cultivation processes. *Renewable and Sustainable Energy Review, 15*(8), 4002–4012. https://doi.org/10.1016/j.rser.2011.07.03162

Razzak, S. A., Hossain, M. M., Lucky, R. A., Bassi, A. S., & de Lasa, H. (2013). Integrated CO_2 capture, wastewater treatment and biofuel production by microalgae culturing—A review. *Renewable and Sustainable Energy Review, 27*, 622–653.

Reymann, T., Kerner, M., & Kümmerer, K. (2020). Assessment of the biotic and abiotic elimination processes of five micropollutants during cultivation of the green microalgae *Acutodesmus obliquus. Biomass Conversion Biorefinery Bioresource Technology Report, 11*, 100512. https://doi.org/10.1016/j.biteb.2020.100512

Rodolfi, L., Chini Zittelli, G., Bassi, N., Padovani, G., Biondi, N., Bonini, G., et al. (2009). Microalgae for oil: Strain selection, induction of lipid synthesis and outdoor mass cultivation in a low-cost photobioreactor. *Biotechnology and Bioengineering, 102*(1), 100–112.

Ruiz-Marin, A., Mendoza-Espinosa, L. G., & Stephenson, T. (2010). Growth and nutrient removal in free and immobilized green algae in batch and semi-continuous cultures treating real wastewater. *Bioresource Technology, 101*(1), 58–64.

Sierra, E., Acién, F., Fernández, J., García, J., González, C., & Molina, E. (2008). Characterization of a flat plate photobioreactor for the production of microalgae. *Chemical Engineering Journal, 138*(1–3), 136–147.

Sivakaminathan, S., Wolf, J., Yarnold, J., Roles, J., Ross, I. L., Stephens, E., Henderson, G., & Hankamer, B. (2020). Light guide systems enhance microalgae production efficiency in outdoor high rate ponds. *Algal Research, 47*, 101846.

Slegers, P., Wijffels, R., Van Straten, G., & Van Boxtel, A. (2011). Design scenarios for flat panel photobioreactors. *Applied Energy, 88*(10), 3342–3353.

Sun, Y., Huang, Y., Liao, Q., Fu, Q., & Zhu, X. (2016). Enhancement of microalgae production by embedding hollow light guides to a flatplate photobioreactor. *Bioresource Technology, 207*, 31–38.
Suparmaniam, U., Lam, M. K., Uemura, Y., Lim, J. W., Lee, K. T., & Shuit, S. H. (2019). Insights into the microalgae cultivation technology and harvesting process for biofuel production: A review. *Renewable and Sustainable Energy Review, 115*, 109361.
Tang, H., Chen, M., Simon Ng, K. Y., & Salley, S. O. (2012). Continuous microalgae cultivation in a photobioreactor. *Biotechnology and Bioengineering, 109*(10), 2468–2474.
Wan, M., Liu, P., Xia, J., Rosenberg, J. N., Oyler, G. A., Betenbaugh, M. J., Nie, Z., & Qiu, G. (2011). The effect of mixotrophy on microalgal growth, lipid content, and expression levels of three pathway genes in *Chlorella sorokiniana. Applied Microbiology and Biotechnology, 91*(3), 835–844.
Xiaogang, H., Jalalah, M., Jingyuan, W., Zheng, Y., Li, X., & Salama, E.S. (2020). Microalgal growth coupled with wastewater treatment in open and closed systems for advanced biofuel generation. *Biomass Conversion and Biorefinery*, 1–20.
Xu, H., Miao, X., & Wu, Q. (2006). High quality biodiesel production from a microalga *Chlorella protothecoides* by heterotrophic growth in fermenters. *Journal of Biotechnology, 126*(4), 499–507.
Xu, X., Gu, X., Wang, Z., Shatner, W., & Wang, Z. (2019). Progress, challenges and solutions of research on photosynthetic carbon sequestration efficiency of microalgae. *Renewable and Sustainable Energy Review, 110*, 65–82.
Zhao, B., & Su, Y. (2014). Process effect of microalgal-carbon dioxide fixation and biomass production: a review. *Renewable and Sustainable Energy Review, 31*, 121–132.
Zhu, L. (2015). Microalgal culture strategies for biofuel production: a review. Biofuels, *Bioproducts and Biorefining, 9*(6), 801–814.
Zeng, X., Danquah, M. K., Chen, X. D., & Lu, Y. (2011). Microalgae bioengineering: from CO_2 fixation to biofuel production. *Renewable and Sustainable Energy Review, 15*(6), 3252–3260. https://doi.org/10.1016/j.rser.2011.04.01457
Zhan, J., Rong, J., & Wang, Q. (2017). Mixotrophic cultivation, a preferable microalgae cultivation mode for biomass/bioenergy production, and bioremediation, advances and prospect. *International Journal of Hydrogen Energy, 42*(12), 8505–8517.
Zhang, J., He, Y., Luo, M., & Chen, F. (2020). Utilization of enzymatic cell disruption hydrolysate of *Chlorella pyrenoidosa* as potential carbon source in algae mixotrophic cultivation. *Algal Research, 45*, 101730.
Zhang, Q., Xue, S., Yan, C., Wu, X., Wen, S., & Cong, W. (2015). Installation of flow deflectors and wing baffles to reduce dead zone and enhance flashing light effect in an open raceway pond. *Bioresource Technology, 198*, 150–156.
Zhang, X.-W., Zhang, Y.-M., & Chen, F. (1999). Application of mathematical models to the determination optimal glucose concentration and light intensity for mixotrophic culture of *Spirulina platensis. Process Biochemistry, 34*(5), 477–481.

Chapter 5
Different Wastewater as Growth Medium

5.1 Diversity of Microalgae in Wastewater

Microalgae are considered as potential source to treat the wastewater and utilize the harvested biomass to produce biofuels (Hussain et al., 2021). The history of the feasible practice of microalgae like *Chlorella* sp., *Dunaliella* to treat wastewater and achieve biomass spans around 75 years (Bhatia et al., 2021). The researchers in several nations like Germany, Mexico, and USA are more oriented toward cost-effective microalgae cultivation and biodiesel production. Several advantages are associated with the cultivation of microalgae in wastewater as

(i) alternative cost-effective cultivating medium;
(ii) rich in nitrogen, carbon, phosphorous;
(iii) production of low-cost biomass for biodiesel production.

5.1.1 Municipal Wastewater

Currently, biological wastewater treatment is majorly focused on MWW such as primary and secondary settlings (Wang et al., 2017a). Studies have shown that various microalgae can successfully remove N and P from the municipal wastewater. MWW is highly composed of nutrients, mostly, nitrogen and phosphorous. The biomass production of 0.73 g/L was observed in microalgae *Chlorella* when cultivated in municipal wastewater (He et al., 2021). In addition, secondary treated wastewater is also used as a cultivation medium for various microalgae. *Chlorosarciopsis* sp. produced 3.14 g/L biomass on primary treated sewage wastewater. Tables 5.1 and 5.2 summarize the composition of various wastewater and impact on biomass and lipid enhancement.

M. P. Rai and S. Vasistha, *Microalgae Biotechnology for Wastewater Treatment, Resource Recovery and Biofuels*, https://doi.org/10.1007/978-3-031-31674-6_5

Table 5.1 Typical characteristics of different wastewater (Sakarika et al., 2020; Aketo et al., 2020; Bhatia et al., 2021)

Properties	MWW	POME	TWW	PWW	OMWW
pH	8.40 ± 0.41	4.30 ± 0.21	7.63 ± 5.34	7.82 ± 1.07	5.13 ± 0.03
Alkalinity (meq L^{-1})	–	–	–	–	0.75 ± 0.10
TS (g/L)	–	–	5.86 ± 8.24	–	112 ± 1
VS (mg/L)	–	–	0.29 ± 0.34	0.22 ± 0.23	82.3 ± 6.8
BOD (g/L)	0.01 ± 0.01	0.35 ± 0.07	–	0.18 ± 0.20	–
COD (g/L)	0.05 ± 0.03	1.74 ± 0.54	0.61 ± 0.54	0.39 ± 0.30	169 ± 9
TN (g/L)	14.8 ± 5.6	–	25.7 ± 35.6	19.6 ± 22.2	735 ± 10
TP (g/L)	2.32 ± 3.19	–	–	–	513 ± 1
TC (g/L)	–	–	–	–	37.5 ± 1.7
Soluble phosphorous (g/L)	–	–	2.04 ± 2.79	–	309 ± 4
Soluble carbohydrate (g/L)	–	–	–	–	26.0 ± 0.3
TSS (g/L)	–	19.6 ± 5.59	0.09 ± 0.09	–	40.6 ± 0.6
Zn (mg/L)	–	–	1.52 ± 1.99	–	–
Mn (mg/L)	–	–	0.03 ± 0.02	–	–
Cd (mg/L)	–	–	0.03 ± 0.03	–	–
Cu (mg/L)	–	–	0.05 ± 0.07	–	–
Pb (mg/L)	–	–	0.09 ± 0.01	–	–
Ni (mg/L)	–	–	0.03 ± 0.01	–	–
As (mg/L)	–	–	<0.001	–	–
Fe (mg/L)	–	–	0.14 ± 0.04	–	–

5.1.2 *Agricultural Wastewater*

Agricultural wastewater composed of high nutrient value as they are high in N and P (Picos-Corrales et al., 2020). The richness of nutrients makes the suitable for microalgae cultivation and attains high biomass productivity (Mat Aron et al., 2021). Some literature conferred that high presence of nutrient might impede the microalgae growth as well as high turbidity present in medium can cause the photolimitation and hence reduce the microalgae growth (Mat Aron et al., 2021). Agriculture wastewater is comprised of dairy, piggery, domestic, etc. To overcome this drawback, the wastewater should be diluted prior to the usage. Khalid et al. (2018) reported on various dilution of agriculture wastewater reported the maximum biomass production.

Table 5.2 Cultivation of microalgae in different waste for nutrient removal, biomass, and lipid production

Type of wastewater	Microalgae	TN% removal	TP% removal	Biomass production (mg/l/d)	Lipid content %	References
Soyabean processing wastewater	*Chlorella pyrenoidosa*	88.8	70.3	640	62.5	(Hongyang et al., 2011)
Dairy wastewater	*Scenedesmus quadricauda*	6.21	89.83	58.75	–	(Daneshvar et al., 2018)
Food wastewater	*Scenedesmus obliquus*	69.5	25.2	68.3	19.4	(Ji et al., 2015)
Dairy wastewater + slaughter house wastewater	*Chlorella vulgaris*	60.5	20.23	432	–	(Lu et al., 2016)
Dairy wastewater	*Acutodesmus dimorphus*	100	100	210	23.8	(Chokshi et al., 2016)
Raw + food processing ADW	*Chlorella sorokiniana*	84.7	54.7	195.8	11.9	(Gupta & Pawar, 2018)
	Scenedesmus oblique	86.9	86.9	149.5	18.39	
	Scenedesmus Abundans	85.3	85.3	165	12.24	
PTWW	*Chlorella vulgraris*	80	100	167	–	(Slade & Bauen, 2013)
STWW	*Chlorella vulgaris*	60.2	35.2	508	26	(AlMomani & Örmeci, 2016)

(continued)

Table 5.2 (continued)

Type of wastewater	Microalgae	TN% removal	TP% removal	Biomass production (mg/l/d)	Lipid content %	References
	Micractinium reisseri	85	96	0.19	30	(Abou-Shanab et al., 2014)
	Chlorella sp.	68	71	–	–	(Nagabalaji et al., 2019)
Mixed piggery wastewater	*Chlorella vulgaris*	100	96	–	–	(Zheng et al., 2018)
Industrial wastewater	*Scenedesmus* Sp.	90	88	–	–	(Ma et al., 2017)
Primary treated sewage	Mixed algae	–	–	25	–	(Kadir et al., 2018)
Municipal wastewater	*Chlamydomon as reinhardtii*	–	–	200	25	(Engin et al., 2018)
Industrial	*Micractinium* sp.	–	–	320	34	
PTWW	*Chlorosarcino psis* sp.	87	82	342	36	(Vasistha et al., 2021)
STWW		85	81	214	29	

5.1.3 *Industrial Wastewater*

Most studies of microalgae cultivation in industrial wastewater are focused on the removal of heavy metal rather than N and P removal (Al-Saydeh et al., 2017). The industrial wastewater mainly comprises of metals like Zn, Mn, Cd, Co, etc. (Plöhn et al., 2021). Industrial wastewater involves palm oil mill effluents, textile wastewater, petroleum wastewater, and olive mill wastewater.

5.1.3.1 Palm Oil Effluent

Malaysia is among largest palm oil manufacturers in the world and generates tons of POME (Zulqarnain et al., 2020). The effluent is discharged into water bodies causing, environmental pollution (Emparan et al., 2020). These effluents are high concentration BOD, COD, N, and P. Cultivation of microalgae in POME proposes an alternative aspects to treat tertiary wastewater and utilizes the nutrients for biomass and lipid production (Ding et al., 2020). *Chlamydomonas incerta* was studied for the highest COD removal of 67% (Kamyab et al., 2015).

5.1.3.2 Textile Wastewater

The organic and inorganic streams of TWW can be acidic or alkaline, depending on the source and composition (El-Kassas & Mohamed, 2014). The microalgae treat the wastewater by degrading the dyes and consume them as N source to cultivate. Some microalgae such as *Anabaena flos-aquae, Nostoc elepsosporum, Nostoc linckia, Anabaena variabilis, and Chlorella vulgaris* reported with the dye removal efficiency of 50–100%. In addition, waste effluents from textile also contain some toxic ions which can impede the aquatic life (Sakarika et al., 2020).

5.1.3.3 Petroleum Wastewater

The wastewater from petroleum industry majorly includes the byproducts, and another waste generated may be spillage from the tanks, improper disposal of PWW (Ashwaniy et al., 2020). The composition present in petroleum byproducts comprises benzene, ethyl benzene, and toluene (Sakarika et al., 2020). Few microalgae such as *Parachlorella kessleri* and *Isochrysis* sp. are able to degrade the benzene, toluene, and ethyl toluene (Takáčová et al., 2015). Whereas crude oil was proved to be toxic for microalgae, concentrations of 30–50% resulted in decreased cell density, during lag phase.

5.1.3.4 Olive Mill Wastewater

Treatment of OMWW through microalgae is tedious in process as they have major portion of polyphenolics and phenols which are toxic for microalgae growth (Malvis et al., 2019). Apart from this, OMWW is high in organic and nutrient content (Di Caprio et al., 2021). At the initials stages, the OMWW first treated with membrane filter, to reduce the concentration of polyphenols (Cicci et al., 2013). As the previous literature discussed, the phenols and polyphenols were removed up to 70% by microalgae *Scenedesmus dimorphus* when treated with ultrafilteration (Cicci et al., 2013). Higher concentration of OMWW was not preferred for cultivation of microalgae, and 50% concentration was seen to increase biomass productivity of cells (Sakarika et al., 2020).

5.1.3.5 Distillery Wastewater

DWW treatment using microalgae serves the dual function as the wastewater forms, distillery unit gets treated and also produces byproducts (Amenorfenyo et al., 2019). Utilizing raw DWW proved to be toxic for microalgae as they are dark in color, the high turbidity can cause photolimitation and do not support the growth of the cell (Chowdhary et al., 2017). There are few studies available in the literature, where the undiluted DWW was treated by microalgae (Soleymani Robati et al., 2021). Sugarcane vinasse was used as the cultivating microalgae such as *Micractinium* sp. ME05, *Arthrospira maxima, and Desmodesmus* sp. and reported the higher biomass production (Santos et al., 2016). The possible reason could be the high nitrogen, and carbon content in sugarcane vinasse has supported the microalgal growth (Santana et al., 2017). Several other DWW involves molasses, anaerobic digested vinasse, beet vinasse, and anaerobic digested vinasse g L^{-1} $NaHCO_3$ (Tasic et al., 2021). Furthermore, in some studies, the process parameter for microalgae cultivation was optimized as many factor influences the microalgae growth such as light, photoperiod, pH, and concentration (De Godoi et al., 2019). In a study, central composite design was used to optimize parameters including temperature, light intensity, and vinasse concentration for microalgae growth.

5.2 Advanced Approaches for Elevating Lipid Accumulation in Microalgae

According to the chosen species, nutrients, environmental factors, and the growth phase during which lipid is taken, microalgae lipid composition and quantity vary. As a conventional method for raising the lipid content of microalgae, inducing microalgal lipid accumulation through manipulation of the nutritional regime and surrounding environmental variables has been extensively investigated (Eldiehy et al., 2022).

The main issues with these methods, however, are their low biomass production and their constrained ability to scale up because they need more money and energy. The major drawbacks of traditional methods have recently been effectively solved by the application of several novel strategies, including combined abiotic stress, two-stage culture, the use of nanoparticles, genetic engineering, and the addition of phytohormones.

5.2.1 Combined Abiotic Stress

This approach combines two or more abiotic stimuli in order to harness their synergistic effects and achieve maximal lipid productivity. Numerous recent researches have shown that combining physical stress with nutritional stress can improve lipid output in microalgae. Under the combined influence of high light and salinity stress conditions, the green alga *Chromochloris zofingiensis* produced the greatest total fatty acid (423.0 mg g^{-1}), a fourfold increase above the control (Kou et al., 2020). The highest lipid productivity (82.0 mg L^{-1} day^{-1}) in *C. vulgaris* was obtained under nitrogen-starve and phosphorus-suffice (N−P+) conditions (Chu et al., 2020), which was 1.5 and 4.6 times higher than that achieved under nitrogen-deficient and phosphorus limited (N−P lim.) conditions and sufficient of phosphorus and nitrogen (N+P+) conditions, respectively.

Dunaliella salina's lipid content was maximized to reach 48.52% when N-limited circumstances, high light intensity, and 10.0% CO_2 were combined in the growing medium. The key advantage of this strategy is that it increases the lipid content without decreasing biomass production by allowing one element to offset the detrimental effects of the other. This method is also simple to apply in the mass manufacturing of biodiesel (Yuan et al., 2019).

5.2.2 Two-Stage Cultivation

This technique uses two sequential, complementary phases that help achieve the highest lipid content without impairing biomass output. By creating the ideal environment for microalgae growth, this technique allows for the first stage's maximal biomass production. The second stage involves the introduction of appropriate stress techniques that cause lipid buildup. Using triethylamine and glycerol, the effectiveness of a two-stage cultivation approach was used to increase cell biomass and lipid accumulation in *Dunaliella tertiolecta*. In this investigation, the first step was completed by adding 1.0 g L^{-1} of glycerol to the culture for 16 days, resulting in the maximum cell biomass (1.30 g L^{-1}) (Liang et al., 2019). The examined strain was grown with 100 ppm triethylamine in the second stage to produce the highest lipid weight content and lipid concentration of 37.70% (DCW) and 383.60 mg L^{-1}, respectively. The novel two-stage culture approach was used to increase the lipid and biomass productivity of *C. vulgaris* (ESP-6) (Hu et al., 2019). Maximum biomass productivity was achieved in this system by growing ESP-6 in the first stage under nutrient-sufficient conditions using the cathode chamber of an airlift type microbial carbon capture cell. The previously acquired biomass was immediately transferred to the second stage to be cultivated under the nutrient-starvation condition utilizing an airlift type photobioreactor in order to improve the lipid content of ESP-6. This method's primary barrier is its species dependence; there is currently no proof that it can be used to all types of microalgal strains. A two-stage cultivation method that increased *C. vulgaris'* areal TAG productivity to 80 tons ha^{-1} $year^{-1}$ had a positive impact on the environment because it reduced the need for fossil fuels and the potential for global warming by 54.0 and 50.0%, respectively, compared to fossil-derived diesel (Adesanya et al., 2014).

5.2.3 Utilization of Nanomaterial's in Nutrient Medium

Nanoparticles are defined as particles with a size between 1 and 100 nm, and they differ from bigger bulk material particles in several ways. These differences lead to the shape, distribution, and size features of materials to diversify (Vasistha et al., 2021a, 2021b). In order to encourage lipid accumulation, nanoparticles have been thoroughly investigated in microalgal cultures. The important microalgal metabolites are either positively or negatively impacted by the type and concentration of nanoparticles used. *Scenedesmus* sp., a green microalga, showed an inhibitory response to TiC and TiO_2 nanoparticles, whereas g-C_3N_4 and SiC demonstrated a positive response to biomass and lipid accumulation. Utilizing 20.0 mg L^{-1} of iron oxide nanoparticles (IONPs), *Chlorella pyrenoidosa's* lipid content and biomass output increased, reaching 13.77 wt% and 2.16 g L^{-1}, respectively (De et al., 2021). The highest lipid content was achieved when the concentration of IONPs was increased to 30 mg L^{-1}, with a maximum lipid content of 16.89 wt% (biomass concentration of 2.01 g L^{-1}). Despite the numerous economic benefits, the main drawbacks of employing nanoparticles are the challenges involved in recycling and reusing them, as well as the hazardous nature of some nanoparticles (Kumar et al., 2021).

5.2.4 Incorporation of Phytohormones

Small chemical compounds called phytohormones, such as auxins, gibberellins, and cytokinins, are naturally occurring in plants and control plant cell activity at very low quantities (Singh et al., 2020). Numerous studies have shown that phytohormones and their derivatives have growth-stimulating effects on microalgae and enhance beneficial metabolites such lipids, carbohydrates, and carotenoids. *Desmodesmus* sp. JS07's biomass and lipid content were increased by 1.96 g L^{-1} and 34.88%, respectively, when indole-3-butyric acid (IBA) was added to the culture medium.

Similarly, when benzylaminopurine (BAP) was utilized, an increase of 1.88 g L^{-1} and 31.84% in the biomass and lipid content was noted. Additionally, IBA (10.0 mg L^{-1}) and BAP (5.0 mg L^{-1}) when combined increased biomass and lipid content to 2.34, 0.03 g L^{-1} and 42.43, 1.88%, respectively (Singh et al., 2020). By adding 5.0 ppm of indoleacetic acid to the growing medium, the microalgae lipid content was increased up to 1.5-fold. In addition, employing a two-phase culture, the synergetic effect of combining 2.0 mg L^{-1} of abscisic acid (ABA) and 2.50 mg L^{-1} of 2,4-dichlorophenoxyacetic acid (2,4-D) increased the total lipid content of Phaeodactylum tricornutum up to 4-fold (Zhang et al., 2020a, 2020b). Nevertheless, using phytohormones to increase the lipid accumulation of microalgae produced excellent results. But further research is necessary because it is unknown how it works.

5.3 Advanced Approaches for Microalgae Cultivation in Wastewater

In general, microalgae can efficiently take their primary nutrients for the synthesis of high-value compounds and/or biodiesel from wastewaters where they grow. However, the high costs for the separation of algae from a variety of aqueous mediums remain important impediments to the wastewater treatment use of microalgae (Xu et al., 2020a, 2020b). Therefore, the development of effective culture methods that are simple to harvest is urgently needed.

5.3.1 Immobilized Microalgae Culture

As is well knowledge, the process of harvesting microalgae requires a lot of time, money, and work. Microalgae cell immobilization is thought to be a workable and efficient way to lighten the load of harvesting. Immobilized culture is a term used to describe batch cultures of algae cells that are directly imbedded in aligned polyurethane foam or adsorbed onto pre-made foam, which is different from a standard liquid suspension culture. The growth doubling time is slower for algal cells embedded in alginate than for loose cells, and the alginate obstructs the photosynthetic process. Microalgae have been cultivated on foam plastics; however, the yield was only approximately 2.5 gm^2 per day (Johnson et al., 2010). The bioavailability of oil-rich *Botryococcus braunii* fixed on calcium alginate gels or agar surface was studied to be lowered by 21% in comparison with a liquid suspension culture. Polyurethane is hazardous when used as an embedding agent. While the metabolic activity and biomass production of algal cells were relatively low, a small number of polymers were less hazardous (Lopez et al., 2009). Immobilization techniques can be applied to microalgae culture, with the limitation of light transmission, if the living cells of algae are to be immobilized. The main benefit of this approach is that cells that are immobilized do not experience drastic changes in their physical and chemical conditions (Cruz et al., 2013). Typically, some algal cells have a predisposition to grow on surfaces by naturally sticking to them. Immobilized microalgae are currently used for energy absorption, metabolite production, culture harvesting, and the elimination of worthless materials from culture media.

Immobilized microalgae applications, in particular, have received more attention in recent years. When several different contaminants need to be treated simultaneously, microalgae immobilization technology is expected to demonstrate a number of benefits. However, the amount of some pollutants eliminated is comparatively low, and the treatment plant's retention period is lengthy, which raises a lot of unforeseen costs. Microalgae production on a big scale for commercial purposes is still not justified, and there are still many technological issues that need to be resolved. The application potential of immobilized microalgae in wastewater treatment facilities will increase as these issues are fixed (Srinuanpan et al., 2019).

5.3.2 *Biofilm-Attached Cultivation of Microalgae*

It is crucial to construct a practical and affordable culture reactor given the high cost of separation and recovery technologies in order to reduce the aforementioned drawbacks in biomass harvesting. It is anticipated that removing problematic contaminants from wastewater using biofilm attachment systems will result in the growth of the desired species of microalgae (Hoh et al., 2016). Algae cells create extracellular polymeric compounds, which can be inoculated onto particular porous surfaces, like a filter membrane or other cotton material, to form biofilms, as demonstrated by biofilm-based microalgal production. The contaminants might be let to remain for a specific period of time by choosing different membrane materials and changing the reactor's tilt relative to the biofilm. Through the membrane materials that are soaked in the culture medium, nutrients and water are supplied to algae cells. Microalgae culture with a biofilm attached is another photoautotrophic method that falls under the category of a photobioreactor. However, the liquid media and the algal cells are mainly separated in biofilm attachment culture, and the predominate microalgal cells adhere to the surface of a unique substratum substance, generating a biofilm (Gross et al., 2016). Biofilms, as opposed to suspension-based photobioreactor, do not have homogenously linked algal biomass and the medium, allowing for more economically advantageous cultivation and harvesting (Cheng et al., 2013).

The biomass productivity can ultimately reach 50–80 gm^2 per day when attached biofilm approach was used to cultivate microalgae (Liu et al., 2013). The biofilm attachment culture may offer a promising means of overcoming the high cost of microalgae production due to its high capacity for biomass synthesis. Additionally, biofilm-attached reactors may be suggested to reduce the light intensity of microalgae by increasing the culture area per unit of incident light area, which may also enhance the function of the light-dark cycle in outdoor cultures where sunlight is at its highest intensity (Wang et al., 2017a, 2017b, 2017c). We are aware that different wastewater types contain a range of organic materials and nutrients. Microalgae culture has been extensively researched, and the outcomes for numerous pilot-scale systems have been published. Active microalgae immobilized in synthetic or natural aggregate objects are the fundamental component of immobilized culture systems used to extract nutrients or other types of contaminants from wastewater. However, the development of coated cells and nutritional absorption might be constrained (Wang et al., 2015). The distinction is that, in order to prevent cell suspension, biofilm connected microalgal culture is based primarily on the traits of algal cells linked to particular safe membrane materials.

However, it requires a lot of energy to separate microalgae from their liquid medium because of their small size, density, and negative charge of the cells. The traditional microalga growing systems, including Open ponds or closed photobioreactors, call for high liquid movement and labor-intensive harvesting techniques. The cutting-edge biofilm attachment culture systems do have the ability to remove contaminants from wastewater in an efficient manner (Zhang et al., 2016). The biofilm-attached culture may effectively remove micropollutants, apply a cost-effective harvest, and increase the productivity of microalgal biomass.

Although Open ponds or closed reactors were the main focus of much of the research on algal mass culture, these conventional suspended production systems have significant drawbacks when compared to biofilm-attached systems. Algal biofilm-membrane photobioreactors, which have been designed for wastewater treatment applications, have increased volumetric microalgae production and nutrients removal. Additionally, by simply adding pesticides or antibiotics, the biofilm-attached culture might achieve the reuse of effluent and limit the contamination from wastewater. In contrast to their culture in Open ponds or closed reactors, this will lessen the direct harm that contaminants do to algae cells. Algae biofilm connected photobioreactors still have certain drawbacks, such as the propensity for water loss through evaporation; as a result, the technology is not yet fully developed. These algal biofilm culture methods still require work, but they show promise in terms of production and performance as a future business (Xu et al., 2020a, 2020b).

5.3.3 Anaerobic Baffled Reactor (ABR)

The third-generation high-rate anaerobic reactor known as the anaerobic baffled reactor, ABR, is very well known for its stable operation without the need for electricity or pumping (also known as energy positive operation) and its significant advantages of rapid biodegradation, low stable sludge yields, excellent process stability under shock loads and affordable construction. Alternatively, they are known as upflow anaerobic sludge blanket reactors (Hahn et al., 2015). Researchers claim that sludge washout, which eventually impairs ABR treatment effectiveness and results in poor effluent quality, is the main downside of ABR. Reactor high-flow speed directly affects sludge washout, whereas low velocity generally solves this issue because higher velocity tends to cause more washout (Garuti et al., 2001). Filter material can be utilized to prevent the washout issue, however, doing so elevate the likelihood of clogging. The elimination of primary clarifiers, along with the reduction of separate digestion and the thickening of primary solids sludge, results in considerable financial and operational savings for wastewater treatment facilities. The anaerobic baffled reactor generates methane at low temperatures. Because the ABR does not need to be treated with energy or chemicals. The operating needs exceed the whole amount of energy potential that was produced.

5.3.4 Hybrid Anaerobic Baffled Reactor (HABR)

Lately, to address the aforementioned issues, a hybrid anaerobic baffled reactor (HABR) with enhanced design concepts and principles has been presented. It consists of a front sedimentation chamber, four conventional baffled chambers, followed by two floated filter media chambers (Metcalf et al., 2003). The acrylic sheet having outside measurements of 90, 20, and 30 cm for length, breadth, and depth, respectively, was used to build these HABRs. The first chamber, named as settling, was twice the subsequent chambers. Each chamber was again divided into two portions by hanging baffles, which separated the chamber in down- and upflow zone (Khalekuzzaman et al., 2018). The bottom of the baffle was slanted at 45°, and the down-to-up flow ratio was 1:4. On the front side of each HABR, in the chamber, which was 20 cm from the base, the sample port was installed. In a study, HAPR-PBR was used as a sustainable method for producing algal biofuels from wastewater. The HABR was capable of eliminating > 90% of COD and TSS from wastewater. *Scenedesmus* and *C. vulgaris* co-cultured algae produced biodiesel with 87.9% dominant FAMEs composition (Khalekuzzaman et al., 2019). HABR offers low operational cost and low sludge production; in addition, they have the ability to separate acidegonosis and methogenesis. However, the reactor takes long startup phase and increases solid loss.

5.3.5 Algal Turf Scrubber

A shallow basin on which an attached algal colony (turf) is growing and through which water is continuously pumped which makes up an algal turf scrubber (ATS). In order for algae to develop and create oxygen and remove nutrients from the water, water is pumped out of the water body and back into the raceway (Adey et al., 2011). Algae purify wastewater by consuming both organic and inorganic substances and producing oxygen during photosynthesis. Algal biomass still contains the nutrients that were extracted from the wastewater, and they can be easily recovered. The system provides rapid growth, increased nutrient removal effectiveness, and rapid O_2 production. It may be used to recover resources from many types of wastewater so there is no need to remove the biomass during production (Chia et al., 2018a, 2018b). ATS can be majorly explored for wastewater remediation. The inoculum, nutrients, and CO_2 that are needed by ATS systems can be easily supplied from the air is the major advantage of the process.

5.3.6 AI-Assisted Microalgae Cultivation

Cultivation of microalgae is the process of transferring microalgae from their natural habitat to a regulated or uncontrolled artificial environment. Light, temperature, pH, nutrients, and the availability of CO_2 through the aeration of microalgal culture are the key environmental factors that determine the growth of microalgae (Teng et al., 2020). The development of methods for the optimization of culture technologies is a result of the efforts to maximize biotechnology production. Statistical techniques, such as response surface methodology, can be used to optimize cultivation operations with various variables (Cao et al., 2020). However, using AI techniques like ANNs and evolutionary algorithms provides powerful tools for simulating nonlinear systems without in-depth system kinetic knowledge. The ANNs were used to conduct investigations on the composition of culture media and how that affected the growth of microalgae. The straightforward hybrid neural network model was developed for fed-batch heterotrophic growing of *Chlorella* species without light. The glucose concentration served as the model's input parameter and its output was a particular growth rate (Figueira et al., 2015). The experimental results were accurately predicted by the hybrid neural network model with only one input (glucose concentration), and the model was deemed adequate for optimizing heterotrophic culture. The multilayered feed-forward model was created for the cultivation medium's optimization to raise the biomass of *C. vulgaris*' lipid content. AI system offers cost-effective system and reduces operational cost; however, there could be chance of error in the process.

5.3.7 Combined Studies

In recent years, a variety of innovative methods have been used to treat various types of wastewaters utilizing microalgae. A novel three-chamber microalgal-bioelectrochemical system was established for clean cultivation of *C. vulgaris* for protein production, treatment of industrial (potato juice) organic effluent, and energy production all at the same time. Three chambers—anode, microalgae, and cathode—were used in this system, and the anode and cathode were filled with industrial organic effluent (Pan et al., 2021). They discovered that increasing current led to increased protein content (320.8–552.1 mg protein g1 biomass) and higher biomass output (0.87–1.11 g L^{-1}) than the control. Results also showed that total Kjeldahl nitrogen (49.8–69.0%) and COD (38.7–66.8%) were successfully removed. In recent study, scientists utilized the MA-MPBR (microalgal-bacterial membrane photobioreactor) to treat high intensity industrial anaerobic digestion effluent, which has the benefit of removing nutrients and organics from sewage in one stage.

In addition, the airlift-type photosynthetic microbial fuel cells were developed to simultaneously recover bioenergy and bioremediate the swine wastewater. Using two-fold dilution of swine wastewater, this system successfully eliminated the COD (96.3%), total organic carbon (95.1%), NH_4^+-N (99.1%), and total phosphorus (98.9%). At a two-fold dilution of swine wastewater, they also reached the greatest 1149 mg L^{-1} d^{-1} CO_2 biofixation rate, 164 mg L^{-1} d^{-1} lipid, and 1.92 kWh m^{-3} net energy output. In a pilot-scale tubular Photobioreactor. The anaerobic food processing wastewater treatment using *C. pyrenoidosa* was studied (Maurya et al., 2022). They observed accelerated development in various seasons, with high biomass production ranging from 1.83 to 2.10 g L^{-1} and fat contents ranging from 8.1 to 15.3%.

The undiluted and unsterilized milk whey processing wastewaters was used to culture *Scenedesmus acuminates* (SA) and a mixed population (PM) primarily composed of *Chlorella, Scenedesmus, and Chlamydomonas* species (Nagarajan et al., 2023). They fed a steady 2.5 L culture volume with a 7-day hydraulic retention period for 70 days using a Plexiglas column PBR. They were successful in removing 93% (SA) and 94% (PM) of COD, 88% (SA) and 90% (PM) of total N, and 69% (SA) and 73% (PM) of total P. The bioremediation of aquaculture effluent using the microalgae *Tetraselmis suecica* in a semi-continuous system was studied where a novel mechanistic model tool BIO ALGAE was used based on Monod growth kinetics to anticipate biomass production and nutrient uptake rates. Both for nutrient intake and biomass production, the model was able to successfully mimic the experimental data pattern (Andreotti et al., 2020).

Despite promising, the utilization of organic carbon for microalgae cultivation poses various problems. Microalgae can only consume few organic molecules for their growth, such as acetate, glycerol, or glucose. Thus, complex organic materials present in IWW cannot be remediated by axenic microalgae culture (Dhandayuthapani et al., 2020). The integration of microalgae production with other organisms that break down organic matter was discovered to be promising and is primarily achievable through two ways. The first method involves a two-step process: in the first, organic compounds in IWW are broken down using anaerobic digestion or activated sludge systems, and in the second, macroalgae will recover nutrients from the liquid phase of digestate or the effluent of the active sludge reactor. The second process to make use of the complex organic compounds in IWW, co-cultures of bacteria or fungi with algae are established. In a co-cultivation situation, bacteria or fungi can oxidize organic material that microalgae cannot consume while generating CO_2 for microalgae development. Microalgae produce oxygen through photosynthesis in the meantime to support aerobic oxidation by bacteria or fungus.

References

Abou-Shanab, R. A. I., El-Dalatony, M. M., EL-Sheekh, M. M., Ji, M.-K., Salama, E.-S., Kabra, A. N., & Jeon, B.-H. (2014). Cultivation of a new microalga, *Micractinium reisseri*, in municipal wastewater for nutrient removal, biomass, lipid, and fatty acid production. *Biotechnology and Bioprocess Engineering, 19*(3), 510–518.

Adesanya, V. O., Cadena, E., Scott, S. A., & Smith, A. G. (2014). Life cycle assessment on microalgal biodiesel production using a hybrid cultivation system. *Bioresource Technology, 163*, 343–355.

Adey, W., Kangas, P., & Mulbry, W. (2011). Algal turf scrubbing: Cleaning surface waters with solar energy while producing a biofuel. *BioScience, 61*, 434–441.

Aketo, T., Hoshikawa, Y., Nojima, D., Yabu, Y., Maeda, Y., Yoshino, T., Takano, H., & Tanaka, T. (2020). Selection and characterization of microalgae with potential for nutrient removal from municipal wastewater and simultaneous lipid production. *Journal of Bioscience and Bioengineering, 129*(5), 565–572.

AlMomani, F. A., & Örmeci, B. (2016). Performance of *Chlorella vulgaris, Neochloris oleoabundans*, and mixed indigenous microalgae for treatment of primary effluent, secondary effluent and centrate. *Ecological Engineering, 95*, 280–289.

Al-Saydeh, S. A., El-Naas, M. H., & Zaidi, S. J. (2017). Copper removal from industrial wastewater: A comprehensive review. *Journal of Industrial and Engineering Chemistry, 56*, 35–44.

Amenorfenyo, D. K., Huang, X., Zhang, Y., Zeng, Q., Zhang, N., Ren, J., & Huang, Q. (2019). Microalgae brewery wastewater treatment: potentials, benefits and the challenges. *International Journal of Environmental Research and Public Health, 16*(11).

Andreotti, V., Solimeno, A., Rossi, S., Ficara, E., Marazzi, F., Mezzanotte, V., & García, J. (2020). Bioremediation of aquaculture wastewater with the microalgae *Tetraselmis suecica*: Semi-continuous experiments, simulation and photo-respirometric tests. *Science of the Total Environment, 738*, 139859.

Ashwaniy, V. R. V., Perumalsamy, M., & Pandian, S. (2020). Enhancing the synergistic interaction of microalgae and bacteria for the reduction of organic compounds in petroleum refinery effluent. *Environmental Technology & Innovation, 19*, 100926.

Bhatia, S. K., Mehariya, S., Bhatia, R. K., Kumar, M., Pugazhendhi, A., Awasthi, M. K., Atabani, A. E., Kumar, G., Kim, W., Seo, S.-O., & Yang, Y.-H. (2021). Wastewater based microalgal biorefinery for bioenergy production: Progress and challenges. *Science of the Total Environment, 751*, 141599.

Cao, M., Kang, J., Gao, Y., Wang, X., Pan, X., Liu, P. (2020). Optimization of cultivation conditions for enhancing biomass, polysaccharide and protein yields of *Chlorella sorokiniana* by response surface methodology.

Cheng, P. F., Ji, B., Gao, L. L., Zhang, W., Wang, J. F., & Liu, T. Z. (2013). The growth, lipid and hydrocarbon production of *Botryococcus braunii* with attached cultivation. *Bioresource Technology, 138*, 95–100.

Chia, S. R., Chew, K. W., Show, P. L., Yap, Y. J., Ong, H. C., Ling, T. C., & Chang, J.-S. (2018a). Analysis of economic and environmental aspects of microalgae biorefinery for biofuels production: A review. *Biotechnology Journal, 13*(6), 1700618.

Chia, S. R., Ong, H. C., Chew, K. W., Show, P. L., Phang, S.-M., Ling, T. C., Nagarajan, D., Lee, D.-J., Chang, J.-S. (2018b). Sustainable approaches for algae utilisation in bioenergy production. *Renewable Energy, 129*, 838–852.

Chokshi, K., Pancha, I., Ghosh, A., & Mishra, S. (2016). Microalgal biomass generation by phycoremediation of dairy industry wastewater: An integrated approach towards sustainable biofuel production. *Bioresource Technology, 221*, 455–460.

Chowdhary, P., Yadav, A., Kaithwas, G., & Bharagava, R. N. (2017). Distillery wastewater: A major source of environmental pollution and its biological treatment for environmental safety BT—Green technologies and environmental sustainability. In R. Singh & S. Kumar (eds.) (pp. 409–435). Springer International Publishing. https://doi.org/10.1007/978-3-319-50654-8_18

Chu, F., Cheng, J., Li, K., Wang, Y., Li, X., & Yang, W. (2020). Enhanced lipid accumulation through a regulated metabolic pathway of phosphorus luxury uptake in the microalga *Chlorella vulgaris* under nitrogen starvation and phosphorus repletion. *ACS Sustainainable Chemistry Engineering, 8*(22), 8137–8147.

Cicci, A., Stoller, M., & Bravi, M. (2013). Microalgal biomass production by using ultra- and nanofiltration membrane fractions of olive mill wastewater. *Water Research, 47*(13), 4710–4718. https://doi.org/10.1016/j.watres.2013.05.030

Cruz, I., Bashan, Y., Hernàndez-Carmona, G., de-Bashan, L. E. (2013). Biological deterioration of alginate beads containing immobilized microalgae and bacteria during tertiary wastewater treatment. *Applied Microbiology and Biotechnoly, 97*, 9847–9858.

Daneshvar, E., Zarrinmehr, M. J., Hashtjin, A. M., Farhadian, O., & Bhatnagar, A. (2018). Versatile applications of freshwater and marine water microalgae in dairy wastewater treatment, lipid extraction and tetracycline biosorption. *Bioresource Technology, 268*, 523–530.

De Godoi, L. A. G., Camiloti, P. R., Bernardes, A. N., Sanchez, B. L. S., Torres, A. P. R., da Conceição Gomes, A., & Botta, L. S. (2019). Seasonal variation of the organic and inorganic composition of sugarcane vinasse: Main implications for its environmental uses. *Environmental Science and Pollution Research, 26*(28), 29267–29282.

Dhandayuthapani, K., Sarumathi, V., Prasad, R., Pancha, I., Gupta, S. K., & Nema, A. K. (2020). An overview of industrial wastewater management using integrated algal technologies. *Algae and Sustainable Technologies*, 141–172.

Di Caprio, F., Tayou Nguemna, L., Stoller, M., Giona, M., & Pagnanelli, F. (2021). Microalgae cultivation by uncoupled nutrient supply in sequencing batch reactor (SBR) integrated with olive mill wastewater treatment. *Chemical Engineering Journal, 410*, 128417.

Ding, G. T., Mohd Yasin, N. H., Takriff, M. S., Kamarudin, K. F., Salihon, J., Yaakob, Z., & Mohd Hakimi, N. I. N. (2020). Phycoremediation of palm oil mill effluent (POME) and CO_2 fixation by locally isolated microalgae: *Chlorella sorokiniana* UKM2, Coelastrella sp. UKM4 and *Chlorella pyrenoidosa* UKM7. *Journal of Water Process Engineering, 35*, 101202.

De Souza Trigueiro, N. S., Gonçalves, B. B., Dias, F. C., de Oliveira Lima, E. C., Rocha, T. L., & Sabóia-Morais, S. M. T. (2021). Co-exposure of iron oxide nanoparticles and glyphosate-based herbicide induces DNA damage and mutagenic effects in the guppy (Poecilia reticulata). *Environmental Toxicology and Pharmacology, 81*, 103521.

Eldiehy, K. S., Bardhan, P., Borah, D., Gohain, M., Rather, M. A., Deka, D. & Mandal, M. (2022). A comprehensive review on microalgal biomass production and processing for biodiesel production. Fuel, *324*, 124773.

El-Kassas, H. Y., & Mohamed, L. A. (2014). Bioremediation of the textile waste effluent by Chlorella vulgaris. *The Egyptian Journal of Aquatic Research, 40*(3), 301–308.

Emparan, Q., Jye, Y. S., Danquah, M. K., & Harun, R. (2020). Cultivation of Nannochloropsis sp. microalgae in palm oil mill effluent (POME) media for phycoremediation and biomass production: Effect of microalgae cells with and without beads. *Journal of Water Process Engineering, 33*, 101043.

Engin, I. K., Cekmecelioglu, D., Yücel, A. M., & Oktem, H. A. (2018). Evaluation of heterotrophic and mixotrophic cultivation of novel Micractinium sp. ME05 on vinasse and its scale up for biodiesel production. *Bioresource Technology, 251*, 128–134.

Figueira, C. E., Moreira, P. F., Giudici, R. (2015). Thermogravimetric analysis of the gasification of microalgae *Chlorella vulgaris. Bioresource Technology.*

Garuti, G., Giordano, A., & Pirozzi, F. (2001). Full-scale ANANOX® system performance. *Water SA, 27*(2), 189–198.

Gross, M., Zhao, X., Mascarenhas, V., & Wen, Z. (2016). Effects of the surface physico-chemical properties and the surface textures on the initial colonization and the attached growth in algal biofilm. *Biotechnology for Biofuels, 9*, 38.

Gupta, S., & Pawar, S. B. (2018). An integrated approach for microalgae cultivation using raw and anaerobic digested wastewaters from food processing industry. *Bioresource Technology, 269*, 571–576.

Hahn, M. J., & Figueroa, L. A. (2015). Pilot scale application of anaerobic baffled reactor for biologically enhanced primary treatment of raw municipal wastewater. *Water Research, 87*, 494–502.

He, Z., Han, W., Jin, W., Yang, J., Gao, S., Li, S., Tu, R., Han, S., Chen, Y., & Zhou, X. (2021). Cultivation of *Scenedesmus obliquus* and *Chlorella pyrenoidosa* in municipal wastewater using monochromatic and white LED as light sources. *Waste and Biomass Valorization.*

Hoh, D., Watson, S., & Kan, E. (2016). Algal biofilm reactors for integrated wastewater treatment and biofuel production: A review. *Chemical Engineering Journal, 287*, 466–473.

Hongyang, S., Yalei, Z., Chunmin, Z., Xuefei, Z., & Jinpeng, L. (2011). Cultivation of *Chlorella pyrenoidosa* in soybean processing wastewater. *Bioresource Technology, 102*(21), 9884–9890.

Hu, X., Liu, B., Deng, Y., Bao, X., Yang, A., & Zhou, J. (2019). A novel two-stage culture strategy used to cultivate *Chlorella vulgaris* for increasing the lipid productivity. *Separation and Purification Technology, 211*, 816–822.

Hussain, F., Shah, S. Z., Ahmad, H., Abubshait, S. A., Abubshait, H. A., Laref, A., Manikandan, A., Kusuma, H. S., & Iqbal, M. (2021). Microalgae an ecofriendly and sustainable wastewater treatment option: Biomass application in biofuel and bio- fertilizer production. A review. *Renewable and Sustainable Energy Reviews, 137*, 110603.

Ji, M.-K., Yun, H.-S., Park, S., Lee, H., Park, Y.-T., Bae, S., Ham, J., & Choi, J. (2015). Effect of food wastewater on biomass production by a green microalga *Scenedesmus obliquus* for bioenergy generation. *Bioresource Technology, 179*, 624–628.

Johnson, M. B., & Wen, Z. (2010). Development of an attached microalgal growth system for biofuel production. *Applied Microbiology and Biotechnology, 85*, 525–534.

Kadir, W. N. A., Lam, M. K., Uemura, Y., Lim, J. W., & Lee, K. T. (2018). Harvesting and pre-treatment of microalgae cultivated in wastewater for biodiesel production: A review. *Energy Conversion and Management, 171*, 1416–1429.

Kamyab, H., Din, M. F. M., Keyvanfar, A., Majid, M. Z. A., Talaiekhozani, A., Shafaghat, A., Lee, C. T., Shiun, L. J., & Ismail, H. H. (2015). Efficiency of microalgae chlamydomonas on the removal of pollutants from palm oil mill effluent (POME). *Energy Procedia, 75*, 2400–2408.

Khalekuzzaman, M., Hasan, M., Haque, R., & Alamgir, M. (2018). Hydrodynamic performance of a hybrid anaerobic baffled reactor (HABR): Effects of number of chambers, hydraulic retention time, and influent temperature. *Water Science Technology, 15*.

Khalekuzzaman, M., Alamgir, M., Islam, M. B., & Hasan, M. (2019). A simplistic approach of algal biofuels production from wastewater using a hybrid anaerobic baffled reactor and photobioreactor (HABR-PBR) system. *PLoS ONE, 14*(12), e0225458.

Khalid, A. A. H., Yaakob, Z., Abdullah, S. R. S., & Takriff, M. S. (2018). Growth improvement and metabolic profiling of native and commercial *Chlorella sorokiniana* strains acclimatized in recycled agricultural wastewater. *Bioresource Technology, 247*, 930–939.

Kou, Y., Liu, M., Sun, P., Dong, Z. & Liu, J. (2020). High light boosts salinity stress-induced biosynthesis of astaxanthin and lipids in the green alga Chromochloris zofingiensis. *Algal Research, 50*, 101976.

Kumar, J. A., Krithiga, T., Manigandan, S., Sathish, S., Renita, A. A., Prakash, P., Prasad, B. N., Kumar, T. P., Rajasimman, M., Hosseini-Bandegharaei, A., & Prabu, D. (2021). A focus to green synthesis of metal/metal based oxide nanoparticles: Various mechanisms and applications towards ecological approach. *Journal of Cleaner Production, 324*, 129198.

Liang, M.-H., Xue, L.-L., & Jiang, J.-G. (2019). Two-stage cultivation of *Dunaliella tertiolecta* with glycerol and triethylamine for lipid accumulation: A viable way to alleviate the inhibitory effect of triethylamine on biomass. *Applied and Environment Microbiology, 85*(4), 1–10.

Liu, T., Wang, J., Hu, Q., Cheng, P., Ji, B., Liu, J., Chen, Y., Zhang, W., Chen, X., Chen, L., Gao, L., Ji, C., & Wang, H. (2013). Attached cultivation technology of microalgae for efficient biomass feedstock production. *Bioresource Technology, 127*.

Lopez, C., Fernandez, F., Sevilla, J., Garcia, M., Moya, B., & Grima, E. (2009). Removal of CO_2 from flue gases coupled to the photosynthetic generation of organic matter by cyanobacteria. *New Biotechnology, 25*, S265.

Lu, Q., Zhou, W., Min, M., Ma, X., Ma, Y., Chen, P., Zheng, H., Doan, Y. T. T., Liu, H., Chen, C., Urriola, P. E., Shurson, G. C., & Ruan, R. (2016). Mitigating ammonia nitrogen deficiency in dairy wastewaters for algae cultivation. *Bioresource Technology, 201*, 33–40.

Ma, C., Wen, H., Xing, D., Pei, X., Zhu, J., Ren, N., & Liu, B. (2017). Molasses wastewater treatment and lipid production at low temperature conditions by a microalgal mutant Scenedesmus sp. Z-4. *Biotechnology for Biofuels, 10*(1), 111.

Malvis, A., Hodaifa, G., Halioui, M., Seyedsalehi, M., & Sánchez, S. (2019). Integrated process for olive oil mill wastewater treatment and its revalorization through the generation of high added value algal biomass. *Water Research, 151*, 332–342.

Mat Aron, N. S., Khoo, K. S., Chew, K. W., Veeramuthu, A., Chang, J.-S., & Show, P. L. (2021). Microalgae cultivation in wastewater and potential processing strategies using solvent and membrane separation technologies. *Journal of Water Process Engineering, 39*, 101701.

Maurya, R., Zhu, X., Valverde-Pérez, B., Kiran, B. R., General, T., Sharma, S., Sharma, A. K., Thomsen, M., Mohan, S. V., Mohanty, K., & Angelidaki, I. (2022). Advances in microalgal research for valorization of industrial wastewater. *Bioresource Technology, 343*, 126128.

Metcalf & Eddy, I. (2003). Wastewater engineering: Treatment and reuse [Internet], 4th ed. revised by G. Tchobanoglous, L. Burton Franklin, S. H. David. Boston: McGraw-Hill.

Nagabalaji, V., Sivasankari, G., Srinivasan, S. V., Suthanthararajan, R., & Ravindranath, E. (2019). Nutrient removal from synthetic and secondary treated sewage and tannery wastewater through phycoremediation. *Environmental Technology, 40*(6), 784–792.

Nagarajan, D., Chen, C. Y., Lee, D. J., & Chang, J. S. (2023). Photobioreactors for microalgae-based wastewater treatment. In *Current developments in biotechnology and bioengineering* (pp. 121–152).

Pan, M., Su, Y., Zhu, X., Pan, G., Zhang, Y., & Angelidaki, I. (2021). Bioelectrochemically assisted sustainable conversion of industrial organic wastewater and clean production of microalgal protein. *Resources, Conservation and Recycling, 168*, 105441.

Picos-Corrales, L. A., Sarmiento-Sánchez, J. I., Ruelas-Leyva, J. P., Crini, G., Hermosillo-Ochoa, E., & Gutierrez-Montes, J. A. (2020). Environment-friendly approach toward the treatment of raw agricultural wastewater and river water via flocculation using chitosan and bean straw flour as bioflocculants. *ACS Omega, 5*(8), 3943–3951.

Plöhn, M., Spain, O., Sirin, S., Silva, M., Escudero-Oñate, C., Ferrando-Climent, L., Allahverdiyeva, Y., & Funk, C. (2021). Wastewater treatment by microalgae. *Physiologia Plantarum.*

Sakarika, M., Koutra, E., Tsafrakidou, P., Terpou, A., & Kornaros, M. (2020). *Chapter 20—Microalgae-based remediation of wastewaters.* In B. P. Yousuf (ed.) (pp. 317–335).

Santos, R., Araujo, O., de Medeiros, J., & Chaloub, R. (2016). Cultivation of Spirulina maxima in medium supplemented with sugarcane vinasse. *Bioresource Technology, 204.*

Santana, H., Cereijo, C. R., Teles, V. C., Nascimento, R. C., Fernandes, M. S., Brunale, P., Campanha, R. C., Soares, I. P., Silva, F. C. P., Sabaini, P. S., Siqueira, F. G., & Brasil, B. S. A. F. (2017). Microalgae cultivation in sugarcane vinasse: Selection, growth and biochemical characterization. *Bioresource Technology, 228*, 133–140.

Singh, J., Jain, D., Agarwal, P., Singh, R. (2020). Auxin and cytokinin synergism augmenting biomass and lipid production in microalgae Desmodesmus sp. JS07. *Process Biochemistry, 95*, 223–234.

Slade, R., & Bauen, A. (2013). Micro-algae cultivation for biofuels: Cost, energy balance, environmental impacts and future prospects. *Biomass and Bioenergy, 53*, 29–38.

Soleymani Robati, S. M., Nosrati, M., Ghanati, F., Hajnowrouzi, A., Grizeau, D., & Dupré, C. (2021). Increase in lipid productivity and photosynthetic activities during distillery wastewater decolorization by *Chlorella vulgaris* cultures. *Applied Microbiology and Biotechnology, 105*(8), 3339–3351.

Srinuanpan, S., Cheirsilp, B., Boonsawang, P., & Prasertsan, P. (2019). Immobilized oleaginous microalgae as effective two-phase purify unit for biogas and anaerobic digester effluent coupling with lipid production. *Bioresource Technology, 281.*

Takáčová, A., Smolinská, M., Semerád, M., & Matúš, P. (2015). Degradation of btex by microalgae *Parachlorella kessleri*. *Petroleum and Coal, 57*, 101–107.

Tasic, M. B., Bonon, A. de J., Rocha Barbosa Schiavon, M. I., Colling Klein, B., Veljković, V. B., & Maciel Filho, R. (2021). Cultivation of *Chlamydomonas reinhardtii* in anaerobically digested vinasse for bioethanol production. *Waste and Biomass Valorization, 12*(2), 857–865.

Teng, S. Y., Yew, G. Y., Sukačová, K., Show, P. L., Máša, V., & Chang, J. S. (2020). Microalgae with artificial intelligence: A digitalized perspective on genetics, systems and products. *Biotechnology Advances, 44*, 107631.

Vasistha, S., Khanra, A., Clifford, M., & Rai, M. P. (2021). Current advances in microalgae harvesting and lipid extraction processes for improved biodiesel production: A review. *Renewable and Sustainable Energy Reviews, 137*, 110498.

Vasistha, S., Khanra, A., Clifford, M., & Rai, M. P. (2021a). Current advances in microalgae harvesting and lipid extraction processes for improved biodiesel production: A review. *Renewable and Sustainable Energy Reviews, 137*, 110498.

Vasistha, S., Khanra, A., & Rai, M. P. (2021b). Influence of microalgae-ZnO nanoparticle association on sewage wastewater towards efficient nutrient removal and improved biodiesel application: An integrated approach. *Journal of Water Process Engineering, 39*, 101711.

Wang, X., Zhao, B., Tang, X., & Yang, X. (2015). Comparison of direct and indirect pyrolysis of micro-algae Isochrysis. *Bioresource Technology, 179*, 58–62.

Wang, J.-H., Zhang, T.-Y., Dao, G.-H., Xu, X.-Q., Wang, X.-X., & Hu, H.-Y. (2017a). Microalgae-based advanced municipal wastewater treatment for reuse in water bodies. *Applied Microbiology and Biotechnology, 101*(7), 2659–2675.

Wang, H., Zhou, W., Shao, H., & Liu, T. (2017b). A comparative analysis of biomass and lipid content in five Tribonema sp. strains at autotrophic, heterotrophic and mixotrophic cultivation. *Algal Research, 24*, 284–289.

Wang, J. F., Liu, W., & Liu, T. Z. (2017c). Biofilm based attached cultivation technology for microalgal biorefineries—A review. *Bioresource Technology, 244*, 1245–1253.

Xu, Z., Wang, H., Cheng, P., Chang, T., Chen, P., Zhou, C., & Ruan, R. (2020a). Development of integrated culture systems and harvesting methods for improved algal biomass productivity and wastewater resource recovery—A review. *Science of the Total Environment, 746*, 141039.

Xu, J., Cheng, J., Xin, K., Xu, J., & Yang, W. (2020b). Strengthening flash light effect with a pond-tubular hybrid photobioreactor to improve microalgal biomass yield. *Bioresource Technology, 318*, 124079.

Yuan, Y., Li, X., & Zhao, Q. (2019). Enhancing growth and lipid productivity in *Dunaliella salina* under high light intensity and nitrogen limited conditions. *Bioresource Technology, 7*, 100211.

Zhang, J., He, Y., Luo, M., & Chen, F. (2020a). Utilization of enzymatic cell disruption hydrolysate of *Chlorella pyrenoidosa* as potential carbon source in algae mixotrophic cultivation. *Algal Research, 45*, 101730.

Zhang, H., Yin, W., Ma, D., Liu, X., Xu, K., & Liu, J. (2020b). Phytohormone supplementation significantly increases fatty acid content of *Phaeodactylum tricornutum* in twophase culture. *Journal of Applied Phycology, 33*, 13.

Zhang, T., Hu, H., Wu, Y., Zhuang, L., Xu, X., Wang, X., & Dao, G. H. (2016). Promising solutions to solve the bottlenecks in the large-scale cultivation of microalgae for biomass/bioenergy production. *Renewable and Sustainable Energy Review, 60*, 1602–1614.

Zheng, H., Liu, M., Lu, Q., Wu, X., Ma, Y., Cheng, Y., Addy, M., Liu, Y., & Ruan, R. (2018). Balancing carbon/nitrogen ratio to improve nutrients removal and algal biomass production in piggery and brewery wastewaters. *Bioresource Technology, 249*, 479–486.

Zulqarnain, Yusoff, M. H., Ayoub, M., Jusoh, N., & Abdullah, A. Z. (2020). The challenges of a biodiesel implementation program in Malaysia. *In Processes, 8*(10).

Chapter 6
Harvesting and Lipid Extraction Techniques of Microalgae in Wastewater

6.1 Harvesting of Microalgae Biomass

Harvesting of microalgae biomass is the major bottleneck in biofuel technology, and it counts for one third of the total cost. Harvesting of microalgae is tedious process due to the small size of the cells and low density of culture suspension. Major applied harvesting technologies are discussed below.

6.1.1 Centrifugation

It is most commonly used method for biomass harvesting, where centrifugal force is applied to concentrate the broth (Laamanen et al., 2016). This method is rapid and devoid of any chemical and bacterial contamination also offers high recovery rate. The maintenance cost and high consumption of energy makes the whole technique a costly process (Mathimani & Mallick, 2018). Hence, for this reason, centrifugation is not preferred in industries or large-scale harvesting. Limitations associated with this technique involve the cell damage due to the generated heat and sheer stress, the cells release the intracellular products (Kotasthane, 2017). It is worthy to note that further downstream processing is required after collective the biomass upon the mixing of intracellular product that also add up the cost. Nevertheless, filtration is another technique for harvesting fragile microalgae cells, described in next section.

6.1.2 Filtration

Current advancement in dewatering of microalgae cells involves membrane filtration to separate microalgae cells which have small size, and the process involves the suspension broth to pass through the filter (Tan et al., 2018). As the membrane

M. P. Rai and S. Vasistha, *Microalgae Biotechnology for Wastewater Treatment, Resource Recovery and Biofuels*, https://doi.org/10.1007/978-3-031-31674-6_6

filter thickens, the flow rate decreases due to clogging, so the replacement of filter is frequently needed to continue the efficiency; hence, this process adds up the cost. In general, the membrane filter is of two types, namely microfiltration and ultrafiltration, having the pore size from 100 to 10,000 nm and 1 to 100 nm, respectively (Barros et al., 2015). Another conventional filtration methods involve microstrainers for larger microalgae cells *Coelastrum proboscideum, Spirulina* having the size more than 70 mm. Different shapes like hollow, tubular, compressed, and spiral are used to prepare membranes which depends on its application (Vasistha et al., 2021). For example, for the harvesting of marine microalgae like *Haslea ostrearia* and *Skeletonema costatum*, the polymer membranes are effective (Suparmaniam et al., 2019).

6.1.3 Gravitational Sedimentation

This process involves the gravity forces for the microalgae settling. It is the most easy and economical process. The sedimentation rate highly depends on the microalgae cell density and the size of the cells, the higher density cells will settle faster than the lighter ones (Menegazzo & Fonseca, 2019). The disadvantage associated with the process involves the lower rate of biomass recovery and time consuming. The longer time for biomass recovery can lead to the deterioration of biomass (Mallick et al., 2016).

6.1.4 Flocculation

It is deliberated as the cost-effective technique for biomass harvesting, and this process can be used for large volume of cultures. In this process, microalgae cells aggregate in presence of flocculants. The mechanism of flocculation works on three forces: charge neutralization, adsorption, and adsorption bridging (Pugazhendhi et al., 2019). The positively charged flocculants are added to the algae culture, absorb negative charge of the cells, and subsequently balance the charge. Thus, the electrostatic repulsion between the particles disappear and hence cells will coagulate (Zhu et al., 2018). Adsorption or electrostatic patch mechanism is where cationic polymers bind with the cells of opposite charge and reverse the charge on the cell surface. This results in patch formation over the boundary of the cells which connect with each other, thus causing flocculation.

Adsorption bridging is another process involved in flocculation where a bridge is formed between two cells with the help of charged polymers that bring the cells together and causes flocculation. Flocculation can be carried out using chemical flocculants or metal flocculants (Li et al., 2020). Alum, ferric chloride, chitosan, cellulose, starch, and guar gum used for microalgae cell harvesting (Blockx et al., 2018).

6.1.4.1 Inorganic Flocculants

Chemical flocculation is the most efficient method for producing bulk microalgal biomass because of its high harvesting flocculation efficiency (FE) and technological economy. With their cationic charge (3+), the flocculants chloride, aluminum, and ferric sulfate are frequently employed for flocculation (Laamanen et al., 2016). Numerous variables, such as pH, electronegativity, ionic charge density, and solubility, have an impact on flocculation efficiency. Microalgal biomass inorganic flocculants are poisonous when utilized in the food sector. For instance, *Nannochlorsarcinopsis oculata* that has been flocculated contains aluminum sulfate; as a result, *Brachionus plicatilis* grows less after feeding than it does when fed *Moringa oleifera* after oil extraction (MOAE) (Noor et al., 2016). Similar to this, after lipid extraction, the harvested biomass residue of *Nannochloropsis salina* with metal salt harvesting has high salt concentrations. Following DAF in water treatment, aluminum and polyaluminum chloride coagulants may efficiently remove different types of microalga in 80% of cases; in particular, alum and ferric chloride ($FeCl_3$) can effectively eliminate *C. vulgaris* (Wang et al., 2011).

In order to produce stable flocs through neutralization, ferric salts, aluminum chloride ($AlCl_3$), and aluminum sulfate $Al_2(SO_4)_3$ are utilized in flotation. However, inorganic coagulants have downsides such as leftover compounds (aluminum and ferric) that might have an adverse effect on the ecology in aquatic environments as well as harmful and non-biodegradable downstream products (Laamanen et al., 2016). A new flocculation method for harvesting *Chlorella* sp. KR-1 was explored where, the coagulant was sulfuric acid (H_2SO_4) and ferric sulfate ($Fe_2(SO_4)_3$). This method can address the reuse of coagulants and biomass pollution, two significant issues (Kim et al., 2017). The pH decrease successfully releases precipitates attached to microalgae, and the residual acidic solution containing recovered ferric ions can be recycled with a harvesting efficiency of greater than 98%. *Scenedesmus obliqus*, a green microalga, produces the maximum FE of 99.5% in 40 min when cultured in municipal wastewater with 150 mg L^{-1} $Fe_2(SO_4)_3$ as a flocculant. Further study is necessary since the mechanism of $Fe_2(SO_4)_3$ for microalga flocculation is not well understood (Abomohra et al., 2018). These salts have limited uses in the production of high-value products and are only beneficial for the treatment of wastewater. This is because recovered biomass may be loaded with metals, which would impair the quality of the product and the downstream process when harvesting microalgae for biofuel.

6.1.4.2 Organic Flocculants

Biodegradable OFs, like chitosan, are plentiful naturally occurring substances with long polymer chains and a positively charged density that make them ideal for charge neutralization and bridging. Chitosan has been extensively researched for use in microalgal harvesting, and it functions well in freshwater at low pH levels of 7.5 and in seawater at high pH levels of > 7.535. Freshwater microalga *C. vulgaris* can be flocculated with chitosan more successfully than with $Al_2(SO_4)_3$ at a dose of 0.25 g L^{-1}, and the medium can be reused for subsequent cultivations with robust growth and lipid content (32.9%) up to (4.6%) (Zhu et al., 2018). Chitosan additionally has no impact on the subsequent microalgae lipid extraction, which makes it superior to metal salts. But the main technical obstacles are the high production costs, the narrow pH range, and the subpar performance of marine algae species. Additionally, this approach only works for green microalgae. For instance, chitosan used to examine the flocculation characteristics of various microalgae and found that it only works for fresh microalgae and performs poorly for marine species. For the purpose of collecting microalgae, buoy-bead flotation has been employed with chitosan treatment as a pre-concentration step. *Chlorella vulgaris* and *Scenesdesmus obliquus* are flocculated using two distinct flocculants, namely inorganic ($FeCl_3$) and organic (chitosan), and their impact on microalgal microsphere aggregates is investigated. The pre-concentration process using $FeCl_3$ results in the release of extracellular organic substances. These organic substances function as polyelectrolytes that are anionic and nonionic and aid in flocculation. There is precipitation when ferric chloride is present. This process involves the entrapment and sweeping down of microalgal cells in the suspension, which results in the formation of flocs via a patching-sweeping mechanism (Zou et al., 2018). In comparison with the microspheres in the solution pre-flocculated with ferric chloride, those in the solution pre-flocculated with chitosan are rougher. The existence of positively charged microspheres in chitosan is confirmed by SEM pictures. Additionally, there are patches of negatively charged microspheres. Microalga-microsphere aggregates were created as a result of these microspheres' reduced repulsion. *Chlorella sorokiniana* is harvested using a brand-new cellulose derivative known as hairy cationic nanocrystalline cellulose (CNCC). This kind of flocculant is coupled to a neutral crystalline shaft by two positively charged ends that are crucial for the creation of flocs. In aggregation, there are two stages. The highest amount of cell area is covered in the first stage, when CNCC becomes adsorbed on the surface of the cells. The second stage, when free spots of cells are covered orthogonally with CNCC (Lopez et al., 2019). The flocculants are categorized into three groups, namely anionic (Magnafloc® 351), cationic (Zetag® 8185, Zetag® 8165, Zetag® 7652, Tanfloc® SG, chitosan, Tanfloc® SH, and Magnafloc® LT22), and nonionic (Zetag® 4120). The best flocculants include Zetag® 8185, Zetag® 8165, and chitosan; at lab and semi-pilot scales, these flocculants have high efficiency reaching 95% between 30 min and 2 h (Muhammad et al., 2021). Additionally, a novel technique for harvesting microalgal biomass using the polymer surfactant aggregate (PSA) approach has been devised. PSA was tested using six different microalgal strains, including *Chlorella* sp. ZTY4, *Scenedesmus* sp. LX1, *Syncohococcus* sp. PCC 7002, *Chlamydomonas reinhardti*, *Microcystis aeruginosa*, and *Selenastrum capricornutum.* Electrostatic interactions and hydrophobic contacts are the two main dominating factors at play in PSA for microalgal recovery. Surfactants with opposing charges enriched, and aggregates developed on the nearby polymer chain. Surfactant aggregates electrostatically bind to microalgae when PSAs develop in solutions. Flocs are created by the binding of

PSA-PSA and PSA-PSA cells (Wu et al., 2018). Five other species had recovery efficiencies of greater than 80%. However, *Scenedesmus* sp. LX1 had a 70% recovery efficiency. The production of positively and negatively charged nanosized PSAs in in-situ culture is responsible for the excellent recovery rates (Pandey et al., 2019). The harvesting of *Chlorella pyrenoidosa* by laboratory-available calcium carbonate ($CaCO_3$) and aluminum sulfate ($Al_2(SO_4)_3$) was studied, and the chemistry of flocculation was examined. A 95% harvesting efficiency for $Al_2\ (SO_4)_3$ was reported. The fundamental benefit of this flocculant is that it preserves cell structures, in contrast to alum, which weakens them, as seen by SEM pictures. $CaCO_3$ slightly alters the biomass, which can then be used for aquaculture, pastoral animal feed, and poultry feed. Furthermore, the effectiveness of harvesting was examined at various pH levels. At a pH of 4.0, which is highly acidic, harvesting effectiveness was high; but, at pH 2.0, it reduced. Possible zeta values can be used to characterize this behavior (Pandey et al., 2019). The isoelectric point is reached at pH 4.0, and when pH drops (from 4.0 to 2.0), a positive zeta potential value (+0.69 mV) is seen. The zeta potential is negative at alkaline pH (Pandey et al., 2019). Cells aggregate and settle through straightforward gravity sedimentation, which leads to the spontaneous aggregation of particles when pH fluctuates at various levels. Only a few types of algae species can be flocculated during this process by altering pH. The freshwater *Ettlia* sp. YC001 algal cells are only mildly flocculated at pH levels under 10 (FE of 11%), but at pH 10.5 and 12.5, respectively, FE increases to 83% and 94% (Yoo et al., 2015). For *C. vulgaris* and *S. obliquus*, various harvesting methods, such as pH-induced flocculation, inorganic flocculation, dissolved air flotation, chitosan, and electro flotation, are examined using lab-scale experiments and large-scale setup (Koley et al., 2017). The organic flocculant is non-toxic and biodegradable but when flocculation takes place at low temperatures, it produces little flocs or produces brittle flocs that can break apart when applied with force.

6.1.4.3 Physical Flocculants

Microalgae are harvested via electrolytic coagulation, which can be applied to a variety of microalgal species and are environmentally beneficial because it does not require the use of chemicals. When electricity is added to the broth, the negatively charged microalgal cells can be separated (Barros et al., 2015). The three steps of electrocoagulation are as follows:

(1) the production of coagulant via electrolytic oxidation;
(2) the dispersal of particle suspension and emulsion breakdown; and
(3) the summation of flocs following destabilization.

Microalgae cells travel toward anodes during this process, where surface charges are neutralized (Chen et al., 2011). As a result, microalgae cells aggregate as seen. Algal cells were extracted in between 80 and 85%. The recovery efficiency of *M. aeruginosa* biomass harvested by electrocoagulation using aluminum and iron electrodes is 100% and 78%, respectively (Gao et al., 2010). Additionally, experimental studies have demonstrated that the lipids and pigments in microalgal biomass are unaffected by electrocoagulation flocculation. The main drawbacks of this method are an elevated medium temperature, cathode maintenance and fouling, mixing effects, pH changes, and a research gap in electric design. Instead of using an Open pond reactor, the electrolytic coagulation process is appropriate for specifically designed photobioreactors (Gerardo et al., 2015). An important component that can considerably lower the amount of energy needed to harvest microalgae is magnetic particles. Iron oxide is used to apply magnetic nanoparticles (MNPs) to the surfaces of microalgae in the presence of an external magnetic field (Wan et al., 2015). The surface properties of microalgae are the most crucial element. Different kinds of magnetic particles have been created for harvesting microalgae. For the collection or separation of microalgae, two common magnetic particles are iron oxide (Fe_3O_4) and yttrium iron oxide ($Y_3Fe_3O_{12}$). Using Fe_3O_4 to harvest *Nannochloropsis oceanica* resulted in a high separation efficiency of more than 90% (Boli et al., 2017). Notably, pH control allows $Y_3Fe_3O_{12}$ to be utilized after being detached from algal aggregates. Fe_3O_4 is easier to extract from flocs than Yttrium iron garnet at high pH levels. *Chlorella* sp. MTFE-7, a marine microalga, was harvested utilizing a unique method that combined magnetic ferrofluid with flocculation at a dosage of 25 mg L^{-1} to achieve 95–100% harvesting efficiency in a short amount of time (1 min). Microalgal flocs are drawn together by self-produced magnetic ferrofluid, which when combined with $FeCl_3$ improves biomass recovery (Ho et al., 2017) Wastewater recycling and microalgal harvesting are both successfully accomplished using an integrated microalgal ferrofluid harvesting dynamic flow-through system. The harvesting cost of US$0.13 per cubic metre was estimated when magnetic particles was explored for fishpond harvesting. Furthermore, utilizing cationic polyacrylamide (CPAM)-Fe_3O_4 particles results in a total harvesting cost of US$2.07 (particle harvesting plus operational costs) for 1 kg of microalgal biomass (Wang et al., 2015). Magnetic separation cannot be scaled up due to the expensive MNPs and specialized equipment needed for recycling nanoparticles.Please check the clarity of the sentence 'Magnetic particles for fishpond...US $0.13 per cubic meter'.The harvesting cost of US$0.13 per cubic metre was estimated when magnetic particles was explored for fishpond harvesting.

6.1.5 *Bioflocculation*

Bioflocculation is the process of producing microalgal flocculation using various microbes and extracellular polymeric substances (EPS). The process of bioflocculation has been used extensively in wastewater treatment. In comparison with other flocculation techniques, it is more affordable and environmentally-friendly (Alam et al., 2016). The majority of current research focuses on flocculation caused by fungus, algae-algae bioflocculation, bacteria-associated flocculation, and plant-based flocculation (Wan et al., 2015). The process offers high chance of recovery and better operation economics but there is a risk of contamination involves in bioflocculation.

6.1.5.1 Bacterial Associated Flocculation

Interestingly, some alga-associated bacteria, namely *Flavobacteriales*, *Burkholderiales*, *Bacillales*, and *Sphingobacteriales,* promote FE (Mathimann et al., 2018). In contrast to an axenic bacterium-free culture of *Ettlia* sp. that had low biomass and low EPS abundance, the axenic bacteria culture of *Ettlia* sp. had 100% FE (Vu et al., 2019). Microbes can be grown separately or in conjunction with microalgae. Additional organic substrates are required in a medium for bacterial culture with microalgae. Algae and bacteria can be co-cultivated in wastewater that is high in organic waste to meet this need. By doing this, wastewater may be cleaned up while also lowering the price of organic substrates. For instance, it has been researched how to improve nutrient removal and microalgal harvesting from seafood wastewater effluent (SWE) using microbial bioflocculation with *C. vulgaris.* The best culture medium for growing and bioflocculating *C. vulgaris* was untreated SWE (Zhou et al., 2015). The interaction of several bacteria on microalgal cell surfaces produced the flocculation activity of 92%. Escherichia coli flocculation activity in medium was 78%. More than 80% of the total suspended salts (TSS) and 78% of the nutrients have been eliminated (Nguyen et al., 2019). The quality of microalgae biomass for food production may be impacted by this procedure' microbial contamination, which is nevertheless suitable for the production of biofuel. Bacterial-based flocculation is free from the chemical contamination of biomass. On the other hand, this process is specific to microalgae strains.

6.1.5.2 Fungal Induced

The majority of eukaryotic algae exist freely in both land and aquatic environments. Generally, lichens frequently form partnerships with fungi and microalgae in nature (Li et al., 2020). In this symbiotic relationship, algae fix CO_2 through photosynthesis and produce organic chemicals to aid filamentous fungi in growing, which turn entangle algae through the formation of hyphae. The generation of cell pellets and the efficient flocculation of phototrophic and heterotrophic species are confirmed by coculturing non-filamentous microalgal cells with filamentous fungi; this impact can improve the long-term viability of the large-scale microalga business. The sizes of the pellets produced by the fungi *Mucor circinelloides*, *Rhizopus oryzae*, and *Pencillium exansum* are 2–5 mm (Chen et al., 2015). *Trichoderma reessei* QM 9414: *Scenedesmus* sp., a pelletized fungal-algal symbiosis system, has been used at a ratio of 1:2. Using secondary effluent from a seafood processing factory, 94% of microalgal cells were extracted in 10 min (Srinuapan et al., 2018). A few fungi species are also useful as feedstock for biodiesel since their lipid concentration exceeds 30% of their total biomass. Algal harvesting is possible with this bioflocculant technique without the use of flocculants (Ummalyma et al., 2017). Within 24 h, a marine species of *Tetraselmis suecuca* co-cultivated with *Aspergillus famigatus* utilizing swine wastewater achieved harvesting efficiencies of over 90%, and symbiotic growth of lipid production and total biomass was observed. *Chlorella pyrenoidosa* and *Asperigillus fumigatus* pellets quickly flocculate (Muradov et al., 2015). Within a short period of time, 99% of the algal cells flocculated (3 h). In the same study, fast flocculation of wastewater-cultivated algae at 3.5 h was found to be 95% effective. This study significantly advanced our understanding of the mechanics underlying the utilization of pelletized F-A for quick bioflocculation (Bhattacharya et al., 2017). The recovery of *Chlorella* sp. cells by an edible fungal strain (*Pleturotus ostreatus*) has been achieved in a novel fungus-assisted harvesting technique. Within 150 min, a maximum harvesting efficiency of 64.86% was attained. The feedstock created with this method can be used to produce feed and food, and the procedure is cheap and flocculant-free (Luo et al., 2019); however, the technique is limited to laboratory studies.

6.1.5.3 Algal Induced Flocculation

Self-flocculation is the process of using spontaneously flocculating microalgae as a source of bioflocculants to harvest non-flocculating microalgae (Alam et al., 2016). Few of the tens of thousands of microalgal species that have been discovered are capable of self-flocculation. Examples of organisms that can self-flocculate include *Ankistrodesmus falcutus*, *Chlorella vulgaris* JSC-7, and *Ettlia texensis* SAG79.80 (Alam et al., 2014; Salim et al., 2011, 2012). Numerous investigations revealed that the mechanisms of bridging and patching both assist the algal–algal-based bioflocculation process. A bridging process is engaged when a vast network of microalgae cells is produced. On the other hand, patching is the mechanism involved if the cells get closer and form more tight connections through EPS by flocculating microalgae. Biochemical analyzes of the self-flocculating microalgae *C. vulgaris* JSC-7 and *S. obliquus* AS-6-1 by Alam et al. (2014) and Guo et al. (2013) revealed that the polysaccharides produced by these species are what cause the flocculation. As a flocculant a green microalga named *E. texensis* SAG79.80, cells flocculate due to glycoproteins as well 85. In the presence of flocculating agents (glycoproteins or polysaccharides) created by microalgal cells, spontaneous microalgal-bioflocculation happens either as a result of the reduction of negative charges through adsorption or possibly as a result of cell bridging through neutralization. To harvest freely suspended microalga *C. vulgaris* CNW1, the self-flocculating microalga *C. vulgaris* JSC-7 is used (Alam et al., 2014). *Scenesdesmus obliquus* AS-6-1 is employed as self-flocculating algae to harvest *C. vulgaris* CNW11 that is suspended freely (Guo et al., 2013). More research is required to fully comprehend the mechanism of self-flocculation and to expand the use of this form of flocculation to collect algal biomass; the two main directions still in the beginning stages are genetic exploitation and habitat isolation. The main benefit of this approach is that algal-based bioflocculation is chemical-free and a sustainable and economical technique of harvesting because it does not require any chemicals (Aguirre et al., 2022).

6.1.5.4 Plant-Based Flocculation

Natural plant-based flocculants have emerged as an alternative to polymeric flocculants, and their applications in the field of wastewater treatment have been increasing given their non-toxicity, biodegradability, and availability from a wide range of renewable resources and the absence of environmental effect, which makes them environmentally friendly (Lee et al., 2014). Many researchers have identified and reported on the use of flocculation mediated by plant-derived biopolymers for wastewater treatment. After oil extraction, the biomass cake from *M. oleifera* exhibits good bioflocculation efficiency for harvesting *Nannochlorosarcinopsis oculate* (Noor et al., 2016). Mung bean protein extracts were found to be an efficient flocculant for *Nannochloropsis* sp. in a different investigation, with a FE of more than 92% at a dosage of 20 mL^{-1} (Kandasamy et al., 2017). Due to their biodegradability, polysaccharides, which are naturally occurring flocculants, are more environmentally benign than inorganic and OFs. The natural biopolymer cationic inulin, derived from plants, was used to gather the freshwater green alga *Botryococcus* sp. At an ideal dosage of 60 mg L^{-1}, this method demonstrated the maximum elimination effectiveness of 88.61% within 15 min (Rahul et al., 2015). Wastewater treatment and microalgae harvesting are both attempted using flocculation with cationic starch as a flocculant. 1 g of *Botryococcus* and *Chlorella pyrenoidosa* biomass are recovered with cationic starch concentrations of 119 and 89 mg, respectively (Hansel et al., 2014). To find the best flocculants, cationic starch was isolated from maize and potatoes; the maximum efficiencies are above 90% and 85%, respectively (Hansel et al., 2014). The process is cost effective, rapid and offers reusability of flocculant; however, it is limited to certain species.

6.1.6 Flotation

Flotation is considered as fast method as compared to sedimentation, in this process, the air bubble passes through the semi-solid suspension, and allowing cells to float. The surface active components get adhered on the surface of the air bubble and result in recovery of cells (Leite et al., 2020). This process is effective for the cells having the diameter from 10 to 500 μm. Dissolved air flotation is another technique in which flocculation and flotation are used for cell harvesting (Zhang & Zhang, 2019). In this process, alum is used as flocculant for the aggregation of cell; simultaneously, the bubbles are provided by an air compressor. The diameter of the bubble generated is directly proportional to the efficacy of cell harvesting. Apart from effective harvesting of biomass, cost is the major hindrance, and the use of compressor at the pressure of 390 KPa adds the cost (Laamanen et al., 2016).

6.1.7 *Electrical-Based Technique*

Electrical-based technique involves electrophoresis and electro-flocculation. In this, the utilization of electricity improves the process and is regarded as the eco-friendly and chemical-free technique (Geada et al., 2018).

In electrophoresis, two metallic rods are used, one of them is anode, and another is cathode; due to the presence of amine and carboxylic group, microalgae is negatively charged; hence, it will move toward positively charged anode (Azmi et al., 2020). The microalgae cells undergo the charge neutralization where cells start to aggregate into flocs, and cells start to settle at the bottom or they float on the surface depending on the density of the cells (Singh & Patidar, 2018).

Another technique for the separation of biomass referred as electro-flocculation, the process implicates the use of electrodes inside the microalgae suspension which releases the metal flocculants, and hence, flocculation occurs. This flocculation follows three stages such as (i) electrolytic oxidation, (ii) destabilization of the cell, and (iii) formation of flocs (Barros et al., 2015).

6.2 Lipid Extraction Techniques

Lipid extraction is an energy-intensive step which limits the sustainability of microalgae bio-oil production. Microalgae are made up of highly complex cell wall, polysaccharides intercalated with protein (de Carvalho et al., 2020). It is not easy to break the cell wall and extract the lipid completely without the application of a large amount of energy. Extraction of lipids using suitable environmentally-friendly solvents is a challenging area of research in biodiesel production (To et al., 2018). There is an urgent need to find alternative solvents which can be used for lipid extraction without harming the environment and health. Various cell disruption methods are introduced for efficient extraction of lipids and categorized as mechanical and non-mechanical (Vasistha et al., 2021).

6.2.1 *Mechanical Pre-treatment Methods*

Different mechanical methods involve mechanical pressing, ball mill, ultrasonic-assisted, microwave, and homogenization. Use of mechanical methods leads to the loss of biomass, and they are less selective toward lipids.

6.2.1.1 Mechanical Pressing

Mechanical pressing involves the screw press, bead milling, and extruder for separation of biomass from oil. The extraction of lipid through mechanical process provides the less chance of contamination, and this technique is often used in industries for the extraction of seed oil (Mubarak et al., 2015). Microalgae *Nannochlorosarcinopsis oculata* attained 51.05% of lipid content with hydraulic press; in addition, the use of liquid nitrogen improved the process by achieving the lipid content of 94%. The use of liquid nitrogen may have freeze the cells and weaken the cell wall, thus releasing the intracellular lipids (Menegazzo & Fonseca, 2019).

6.2.1.2 Ball Mill

Ball mill comprises of rotating cylinder which act as grinder and made up of metallic beads or quartz (Steriti et al., 2014). The speed of rotation directly influences the cell damage, and high speed offers more damage to the cell. The beads damage cell wall so rapidly and can be applied to wet or dry cell biomass (Patel et al., 2018). There are various factors which influence disruption rate of microalgae cells, including shape of the container, size, stirring speed (Bharte & Desai, 2021). The simple mechanism and speedy process convey the major advantage to the method. The lipid extraction using ball mill is selective to some microalgae like *Botryococcus* sp., *Scenedesmus* sp. (Amorim et al., 2020).

6.2.1.3 High Pressure Homogenizers

This method is also known as French press; in this technique, the biomass is pushed in to the narrow tube through the hydraulic shear force (Carullo et al., 2018). This energy-intensive process generates the heat that may degrade the biomass. The efficacy of high pressure homogenizer varies from species to species, for example, *Chlorococcum* sp. offers 73.8% of cell disruption; on other hand, the microalgae *Chlorella* showed only 16% cell damage (dos Santos et al., 2015; Halim et al., 2012).

Another technique associated with cell disruption known as depressurization where the diffusible gas carbon dioxide penetrates inside the cell wall (Goh et al., 2019). The sharp gradient causes the gas to expand and causes the cell disruption. This method does not involve any toxic solvent and take less time for lipid extraction (Menegazzo & Fonseca, 2019). Researchers found that the time for lipid extraction was reduced to 5 times lesser in case of *Chlorococcum* sp. (Halim et al., 2011).

6.2.1.4 Ultrasonication-Assisted Method

During this process, the ultrasonic waves are generated in the suspension broth causes the low and high fluctuating pressure to result in the formation of microbubbles, which collapse due to high pressure that induces cavitation (Fattah et al., 2020). Hence, the cell structure is ruptured, and lipid is released. Cavitation are more penetrating at the low frequency (18–40 kHz) rather than high frequency (400–800 kHz). Some literature reported that the duration for biodiesel production was decreased by 23 h with 95% yield (Lee et al., 2021). Whereas other studies deliberated that quality of lipid was compromised when using sonication method due to lipid oxidation. This process is suitable for the microalgae which have rigid cell wall like *Chlorella* sp., *Scenedesmus* sp. (Sati et al., 2019).

6.2.1.5 Microwave-Assisted Method

Microwave-assisted methods generate radiations with frequency of 0.3–300 GHz which breaks the cell wall and allow the release of lipids (Kapoore et al., 2018). The method offers the high lipid yield and reduces the duration of the extraction. The energy transfer rate is comparatively high than sonication and thus provides more effective heating (Fazril et al., 2020). However, due to the high temperature, the quality of lipid may get retarded in process. Microalgae like *Chlorella vulgaris, Botryococcus* sp. best results were achieved, using microwave (Harris et al., 2018).

6.2.2 Non-mechanical Pre-treatment Methods

Apart from various mechanical methods introduced for lipid extraction, other methods like chemicals or solvent are efficient for cell wall disruption. The efficiency of lipid extraction also depends on the nature of the solvent, i.e., polar or non-polar. Currently, polar solvents are used in combination with non-polar solvent for complete removal of lipid.

6.2.2.1 Solvent-Based Extraction

The principle for solvent-based extracted state “like dissolve like”. The solvent should have high affinity toward lipids specially TAGs and have low volatile nature for better extraction (Table 6.1).

Solvents, for example, chloroform, methanol, hexane, benzene, di ethyl ether, and toluene, are used for extraction of lipid (Bundhoo, 2018). The most preferred conventional method Bligh and Dyer used for lipid extraction uses chloroform and methanol in the ration of 2:1, and water as co-solvent where the biphasic system occurs and lipid is dissolved in chloroform (Bligh & Dyer, 1959). Although, some

Table 6.1 Target component by solvents for lipid extraction (Lee et al., 2021; Menegazzo & Fonseca, 2019)

Solvents	Compound extracted
Methanol	Phospholipids, glycolipids
Ethanol	Phospholipids, glycolipids
Acetone	Diacylglycerols, sulfolipids
Chloroform	Hydrocarbons, pigment, TAGs, waxes
Hexane	Hydrocarbons, TAGs

solvents are toxic to environment such as chloroform and methanol and also dissolve pigment (Chlorophyll) along with lipid. *Botrycoccus braunii* biomass was explored for the extraction of lipid using five different organic solvent chloroform: methanol (2:1, v/v), dichloromethane: methanol (1:1, v/v), dichloromethane: ethanol (1:1, v/v), acetone: dichloromethane (1:1, v/v) and hexane: isopropanol (3:2, v/v) (Japar et al., 2017; Kumar et al., 2019; Shin et al., 2018). The maximum yield was observed in mixture of chloroform: methanol. Some other solvents like methylene chloride and methanol were used for extraction neutral lipid from *C. vulgaris.*

6.2.2.2 Supercritical CO_2 Extraction

It is considered the suitable solvent for the extraction as the pressure and temperature can be varied according to the conditions (Taher et al., 2020). Supercritical CO_2 preferred for the critical solvents by providing the pressure of 7.4 MPa and temperature of 31.1 °C. The lipid yield is inversely proportional to pressure and temperature applied (Patel et al., 2020). The optimized best condition (35 MPa, 40 °C) for lipid extraction from microalgae *Shizochytrium limacinum* where ethanol was used as co-solvent (Mubarak et al., 2015). Another report obtained the highest lipid content in *Chlorella* sp. with 40 MPa pressure and temperature of 32 °C (Hu et al., 2007). Santana et al. concluded the increase in pressure and density of CO_2 and upsurge the unsaturated percentage in the lipid (Santana et al., 2012). TAGs yield was increase when ethanol is used as co-solvent in *Nanochlorosarcinopsis sp which is oculata* with 35 MPa and 50 °C (Nobre et al., 2013).

6.2.2.3 Ionic Liquids

They are non-aqueous solution made of cation and anion. Ionic liquids are preferred for lipid extraction due to their non-volatility and thermally stability nature, and they remain in liquid state over a wide range of temperature (0–140 °C) (Zhou et al., 2019). Regardless of their potential properties, only few microalgae are studies with ionic liquids for the lipid extraction. While using 1-butyl-3-methylimidazolium for lipid extraction, two separate layers of hydrophobic and hydrophillic were formed (Kumar et al., 2017). Due to gravity, hydrophobic lipid layer moves to upper phase. Ionic liquids like [$MeSO_4$]·[Bmim] [$MeSO_4$] were combined with microwave technique for the better extraction of lipid from *Chlorella vulgaris*. The extracted lipid was 1.6 folds high as compared with Bligh and Dyer method. Utilization of ionic liquids is limited due to the lack understanding cationic and anionic properties of salts (Krishnan et al., 2020).

6.2.2.4 Green Solvents

Inexpensive low molecular weight polymers, deep eutectic solvents, fluorous solvents, and supramolecular solvents are introduced as green solvent. They are regarded as green due their non-toxic, inexpensive, and eco-friendly nature (Jeevan Kumar et al., 2017). Polymers like ethylene glycols and propylene glycols are non-hazardous due to low vapor pressure. Furthermore, deep eutectic solvents are composed two or more compounds such as ammonium salt and metal chloride and ammonium salt and metal chloride hydrate (de Jesus et al., 2019). They are more versatile solvents as they have low melting point, varied polarity range, biodegradable. Fluorous solvents include perfluoro cyclohexane having aqueous and non-aqueous phase. Studies discussed that they can be alternative to for extraction of essential oils. Supramolecular solvents are considered vital for the extraction of polar and non-polar compounds (Vasistha et al., 2021). They are amphiphiles, nanostructured liquids used in the extraction process of bioactive compounds, dyes, and surfactants. In addition, non-explosive nature of supramolecular solvents makes them suitable for green solvent (Ali et al., 2021).

6.2.2.5 Deep Eutectic Solvents (DESs)

DESs are typically made up of non-toxic ingredients that are found in foods naturally and can be employed directly in food compositions. Better selectivity and yield were demonstrated by DESs that pre-treated microalgal biomass with microwave radiation, followed by extractions utilizing dimethyl carbonate (DMC) and supercritical (SC) CO_2. The microalgal biomass pre-treatment with DESs significantly improves the primary extraction procedure. The study's findings indicated that algal lipids recovered relatively quickly. The recovery rates of DESs based on ethylene glycol, acetamide, or ChCl with oxalic acid were about 80%, 67%, and 75%, respectively. The research demonstrates that DESs can break down cell walls and improve extraction performance. DESs-based extraction techniques are typically constrained by viscosity. Dai et al. (2013) proposed a straightforward fix: liquefying the DESs (by adding water in the right proportion) prior to application and developing a pseudo-ternary DESs system with water as the third component. Increased water concentration significantly reduces the viscosity of DES, since a solvent's viscosity is solely dependent on the strength of its H_2 bonds. The higher H_2 bonds increase the solvent's viscosity and decrease the solubility of the solutes (Dai et al., 2013). The thick cell walls that surround microalgae cells are made up of two primary layers: an outer layer that is more densely packed with electrons than the inner layer. This is caused by the existence of algaenan, a biopolymer with a high level of resistance that makes lysis challenging and enables these organisms to withstand chemical and weak acid or base attacks (Mehariya et al., 2020). Hence due to high viscosity of DES, rupture of microalgae cells becomes feasible.

6.2.2.6 Three Phase Partitioning

Three phases, including oil in the top organic phase, proteins in the middle precipitate phase, and some hydrophilic molecules like polysaccharides in the bottom phase, simultaneously develop in TPP. Incubation and extraction steps are included in the most typical TPP method (Liang et al., 2022; Tan et al., 2015). Incubation involves combining a powdered source with a salt solution for a predetermined amount of time, and extraction involves adding t-butanol slowly. In other circumstances, a powdered source, salt, and t-butanol are all introduced to aqueous media at the same time and then vigorously mixed. After the mixing procedure, the three phases were separated using a centrifugal or gravimetric technique. The upper phase is made up of oil dissolved in t-butanol, the intermediate dense phase is proteins precipitated with salts, and the final phase is an aqueous phase made up of water-soluble bio molecules like carbohydrates. Oil was extracted from microalgae using a TPP system based on potassium hydrogen phosphate and ethanol, and the yield was observed to be 69%. Compared to ammonium sulfate, dipotassium hydrogen phosphate (DKP) is a weak protein-precipitating agent hence other researchers' choice for TPP-based oil extraction. For the TPP process of oil extraction, researchers investigated a wide range of organic solvents, including methanol, ethanol, n-propanol, isopropanol, 1-octanol, and t-butanol (Panadare et al., 2017). Three phase division was shown to be a useful technique for extracting protein when carried out at room temperature, whereas triglycerides are dissolved at higher temperature. Many extraction procedures use a variety of techniques to shorten their duration, including heating, mixing, the use of continuous fresh solvent, smaller particle size, etc. To create three phases, TPP mixes well-buffered aqueous salts with an equivalent amount of water-miscible aliphatic alcohol. None of the alcohols, including methanol, ethanol, propanol, and t-butanol, are miscible with ammonium sulfate ($(NH_4)_2SO_4$) in aqueous solutions. As a result, two liquid phases will form, with the salty water remaining below and the alcohol floating to the top. Aqueous solution will then form a third phase in between the two phases, forming a three-phase solution (Chew et al., 2019). Due to its promising separation efficiency during the extraction, the TPP concept resembles a cross between salting out and alcohol precipitation processes.

6.2.2.7 Surfactant Assisted

The ability of surfactants, which have a hydrophobic tail and a hydrophilic head, to interact with cell membranes and break down cells is well established. Commercial surfactants come in a variety of forms, including cationic, anionic, zwitterionic, and nonionic. The membranous components of microalgae can be solubilized by cationic surfactants like cetyltrimethylammonium bromide. Lai et al. (2016) examined the effectiveness of different surfactant families in destroying Scenedesmus's dry and wet biomass. The optimal FAMEs recovery was 160 times greater in the cells treated with the inner salt surfactants 3-(decyldimethylammonio)-propanesulfonate and myristyltrimethylammonium bromide (Lai et al. 2016). Additionally, Wu et al. (2018) looked at the possibilities of co-solvents include external and algal-based surfactants for the extraction of lipids. The outcomes revealed that the combination of using an algal-based surfactant and hexane produced the elevated 78.8% lipid yield, and 78.2% yields were obtained.

6.2.2.8 Nanomaterials

As is well known, nanoparticles possess specific qualities that make them suitable for a wide range of scientific applications. For the treatment of wastewater, metallic nanoparticles have a photocatalytic and antibiofilm impact. Additionally, they possess the photothermal property, which causes heat to dissipate surrounding the particles when light is absorbed. A study uses nanoparticles to show the molecular delivery system in cells without causing cell damage. The photothermal impact of gold nanoparticles can assist in cutting-edge medical treatment approaches for terrible ailments including cancer treatment and biopsy. Metal nanoparticles are only used here, in relation to biofuels, to improve lipids, biomass, and act as a catalyst in transesterification reactions. Using wastewater as a source of nitrogen and phosphorus can act as a link between nanoparticles and algae for the creation of biofuel. Due to the lack of freshwater use during the biofuel manufacturing process and the use of nanoparticles produced from biological sources, the procedure is more environmentally-friendly and sustainable. Nanoparticle engineering provides platform to solve various challenges in microalgae biofuel technology (Fu et al., 2021). In a nutshell, nanoparticle engineering is a study of nanometer size material or the development of artificial nanomaterial (Chintagunta et al., 2021). The unique capability of nanoparticle makes them efficient for various applications including food industry, bioremediation, cosmetics, and drug delivery (Vargas-Estrada et al., 2020). They are stable in nature and have large surface area. Nanomaterials have the potential to upsurge the intracellular lipid in microalgae, and they act as catalyst (Khanra et al., 2020). In addition, nanomaterials can also reutilized for cultivation which influence the economic aspect of the process. Nanomaterials like silver NPs, Fe_3O_4, silica-based NPs, amino clay NPs, and NiO NPs are effective for microalgae lipid extraction (Singh et al., 2020). Huang and Kim stated that lipid extraction efficacy of 91% was achieved for wet microalgae biomass using NiO NPs (Huang & Kim, 2016). The dosage of nanoparticle plays the significant role in lipid extraction. Silica NPs are more used for separation of high-value compounds, like lipids. They have high porosity which makes them efficient for lipid extraction. NPs of 10 nm pore size are capable for capturing free fatty acids. The idea that nanoparticles interact with microalgae to ooze lipids out of the cells is known as the "milking microalgae for lipids" concept. Due to their photothermal characteristics, the surface-adhered nanoparticles can be excited via photonics, producing a localized heating effect. The extraction of high and low value-added compounds without cell death is possible because to the creation of nanopores into the cells by the localized heat dissipation to the cell surface. By making pores on the surface of the cells, electroporation techniques can disturb the cells, but because of their tremendous power, the cells typically do not survive. On the other hand, the proposed method implies that another cycle of lipid extraction is feasible if the cells are provided with the right nutrients and growth circumstances. Future cell harvesting techniques may be made possible by the particles' localized heating action, which is photothermal-based. It will reduce the expense of mechanical drying and cell collection, two procedures that consume a large amount of energy in any biorefinery. This will enable a continuous lipid-producing system by easing the additional strain of inoculating the fresh microalgae culture at each cycle. Various techniques for harvesting and its effect on microalgae lipid have been discussed in Table 6.2.

6.2.2.9 Water Plasma Enhanced Method

The non-thermal plasma approach is a cutting-edge technology because of its distinctive benefits, including its quick, affordable, and convenient operation. Since the 1920s, there has been extensive research on the plasma principle (Wang et al., 2021a, 2021b). Gas molecules smash with high-energy electrons after a particular electric field is activated, turning the gas into plasma. An earlier study revealed that non-thermal-plasma method treatment is frequently employed to change the surface characteristics of food and materials connected to food in order to enhance mass transfer (Liang et al., 2022). Exploiting the technique for cell disruption using the non-thermal plasma strategy in a water medium offers the chance of simultaneous intracellular lipid extraction with less energy use because the drying process of microalgal biomass is avoided in the extraction of lipids from microalgae with high water content. This technique might lead to effective and affordable lipid extraction (Alam et al., 2019).

6.2.2.10 Pulse Microwave Method

Pulsed microwave radiation has been suggested as a substitute for continuous microwave radiation for a variety of uses, such as the manufacture of biodiesel from microalgae (Prommuak et al., 2016). The microalgae *Auxenochlorella protothecoides* can be effectively pre-treated with PMW to remove any water-soluble materials as well as to undertake solvent extraction of the lipids. When PMW treatment is applied to a microalgae suspension concentrated at 100 g/L and with a flow rate of 0.02 mL/s, pulses of 200 ms duration and 2.8 kW peak power enabled to achieve lipid yield of 37.29% dry weight, i.e., extraction of almost 90% of the total lipid content, with an energy consumption not exceeding 2.53 MJ/kgDW (Zhang et al., 2022).

Table 6.2 Summary of various techniques used for harvesting and lipid extraction

Microalgae	Technique	Harvesting efficiency (%)	Lipid content (%)	References
Shizochytrium limacinum	Supercritical (CO_2 and ethanol)	–	33.9	Tang et al. (2011)
Synechocystis	Ethanol	–	48	Sheng et al. (2011)
	Flocculant	≤ 78	–	Barros et al. (2015)
Anabaena sp.	Aluminum sulfate	≤ 95	–	Lemos et al. (2016)
Synechocystis PCC 6803	Chloroform/methanol/water	–	42	Mubarak et al. (2015)
Chlorella sorokiniana	$FeCl_3$	–	66	Mallick et al. (2016)
Chlamydomonas sp.	Cationic gaur gum	–	92	Banerjee et al. (2014)
Monoraphidium contortum	Nb_2O_5	–	94.27 (FAME)	Nguyen et al. (2019)
Spirulina	Zn-Mg NPs	–	37.1	Egesa et al. (2018)
Scenedesmus sp.	SiC	–	40.26	Ren et al. (2020)
Scenedesmus obliquus	Fe_2SO_3	10	–	He et al. (2017)
Acutodesmus obliquus	Ethanol–hexane (1:2)	–	92	Hurtado et al. (2018)
Chlorella sp.	Methanol–ethyl acetate (2:1)	–	18.1	Wang et al. (2021a, 2021b)
Scenedesmus sp.	Hexane	–	16.3	Shin et al. (2014)
Chlorella	Ag	87	–	Romero et al. (2020)
Chlorella	Fe_2O_3	92.4	–	Hena et al. (2016)
Scenedesmus sp.	Fe_3O_4	> 80	–	Ge et al. (2015)
Chlorella vulgaris	Fe-MNP	98.6	–	Almomani and Bohsale (2021)
Desmodesmus subspicatus	Nzvi	–	58.33	Pádrová et al. (2015)
Scenedesmus obliquus	MgO	–	<50	He et al. (2017)
	CNTs	–	<2	

(continued)

Table 6.2 (continued)

Microalgae	Technique	Harvesting efficiency (%)	Lipid content (%)	References
Chlorella sp.	MgAc	–	25	Kim et al. (2016)
Botryococcus braunii	CO_2 methanol	24	–	Paudel et al. (2015)
Nanochlorosarcinopsis sp is	Organosilicone	–	85.95	Sun et al. (2018)
Scenedesmus dimorphus	PEI-Fe_3O_4	> 80	–	Ge et al. (2015)
	Graphene oxide	73.1	–	
Chlorella pyrenoidosa	Sulfonated graphene	48.6	–	Cheng et al. (2017)
Chlorella vulgaris	SrO-C dots	–	98.7	Guldhe et al. (2017)
Chlorella sp.	Centrifugation (4000 rpm)	100	–	Ahmad et al. (2014)
Nannochloropsis sp	Nanochitosan	–	80	Farid et al. (2013)
Chlorella vulgaris	Ultrasound (600 W)	–	14	
Tolypothrix sp.	Ball mill (3500 rpm)	–	28	Menegazzo and Fonseca (2019)
Botryococcus sp.	Microwave (2450 MHz)	–	28.5	
Chlorella sp.	$Fe_2(SO_4)_3$	90	–	Kim et al. (2017)
Phaeodactylum tricornutum	PAC	66.6	–	Suparmaniam et al. (2019)
Schizochytrium sp.	Ultrasound (20 kHz)	–	31.4	Byreddy et al. (2015)
Chlorococcum sp.	Ti	95	–	Khanra et al. (2020)
Chlorella sp.	Fe_3O_4	95	–	Liu et al. (2020a, 2020b)
Scenedesmus sp.	Fe_3O_4	68	–	Abo Markeb et al. (2019)

References

Abo Markeb, A., Llimós-Turet, J., Ferrer, I., Blánquez, P., Alonso, A., Sánchez, A., Moral-Vico, J., & Font, X. (2019). The use of magnetic iron oxide based nanoparticles to improve microalgae harvesting in real wastewater. *Water Research, 159*, 490–500.

Abomohra, A. E.-F., Jin, W., Sagar, V., & Ismail, G. A. (2018). Optimization of chemical flocculation of *Scenedesmus obliquus* grown on municipal wastewater for improved biodiesel recovery. *Renewable Energy, 115*, 880–886.

Aguirre, A. M., Bassi, A., & Saxena P. (2022). Engineering challenges in biodiesel production from microalgae. *Critical reviews in biotechnology, 33*, 293.
Ahmad, A. L., Yasin, N. H. M., Derek, C. J. C., & Lim, J. K. (2014). Comparison of harvesting methods for microalgae *Chlorella* sp. and its potential use as a biodiesel feedstock. *Environmental Technology, 35*(17), 2244–2253.
Alam, M. A., Muhammad, G., Rehman, A., Russel, M., Shah, M., & Wang, Z. (2019). Standard techniques and methods for isolating, selecting and monitoring the growth of microalgal strain. *Microalgae Biotechnology for Development of Biofuel and Wastewater Treatment*, 75–93.
Alam, M. A., Vandamme, D., Chun, W., Zhao, X., Foubert, I., Wang, Z., et al. (2016). Bioflocculation as an innovative harvesting strategy for microalgae. *Reviews in Environmental Science & Biotechnology, 15*, 1–11.
Alam, M. A., Wan, C., Guo, S. L., Zhao, X. Q., Huang, Z. Y., Yang, Y. L., et al. (2014). Characterization of the flocculating agent from the spontaneously flocculating microalga Chlorella vulgaris JSC-7. *Journal of Bioscience and Bioengineering, 118*, 29.
Ali, Z., Subeshan, B., Alam, M. A., Asmatulu, E., & Xu, J. (2021). Recent progress in extraction/transesterification techniques for the recovery of oil from algae biomass. *Biomass Conversion and Biorefinery*. https://doi.org/10.1007/s13399-021-01326-y
Almomani, F., & Bohsale, R. R. (2021). Bio-sorption of toxic metals from industrial wastewater by algae strains Spirulina platensis and Chlorella vulgaris: Application of isotherm, kinetic models and process optimization. *Science of the Total Environment, 755*, 142654.
Amorim, M. L., Soares, J., Vieira, B. B., Batista-Silva, W., & Martins, M. A. (2020). Extraction of proteins from the microalga *Scenedesmus obliquus* BR003 followed by lipid extraction of the wet deproteinized biomass using hexane and ethyl acetate. *Bioresource Technology, 307*, 123190.
Banerjee, C., Ghosh, S., Sen, G., Mishra, S., Shukla, P., & Bandopadhyay, R. (2014). Study of algal biomass harvesting through cationic cassia gum, a natural plant based biopolymer. *Bioresource Technology, 151*, 6–11.
Barros, A. I., Gonçalves, A. L., Simões, M., & Pires, J. C. M. (2015). Harvesting techniques applied to microalgae: A review. *Renewable and Sustainable Energy Reviews, 41*, 1489–1500.
Bharte, S., & Desai, K. (2021). Techniques for harvesting, cell disruption and lipid extraction of microalgae for biofuel production. *Biofuels, 12*(3), 285–305.
Bhattacharya, A., Mathur, M., Kumar, P., Prajapati, S. K., & Malik, A. (2017). A rapid method for fungal assisted algal flocculation: critical parameters & mechanism insights. *Algal Research, 21*, 42–51.
bin Azmi, A. A., Sankaran, R., Show, P. L., Ling, T. C., Tao, Y., Munawaroh, H. S. H., Kong, P. S., Lee, D.-J., & Chang, J.-S. (2020). Current application of electrical pre-treatment for enhanced microalgal biomolecules extraction. *Bioresource Technology, 302*, 122874.
Bligh, E. G., & Dyer, W. J. (1959). *Canadian Journal of Biochemistry and Physiology, 37*(8).
Blockx, J., Verfaillie, A., Thielemans, W., & Muylaert, K. (2018). Unravelling the mechanism of chitosan-driven flocculation of microalgae in seawater as a function of pH. *ACS Sustainable Chemistry & Engineering, 6*(9), 11273–11279.
Boli, E., Savvidou, M., Logothetis, D., Louli, V., Pappa, G., Voutsas, E., et al. (2017). Magnetic harvesting of marine algae *Nannochloropsis oceanica. Separation Science and Technology.*
Bundhoo, Z. M. A. (2018). Microwave-assisted conversion of biomass and waste materials to biofuels. *Renewable and Sustainable Energy Reviews, 82*, 1149–1177.
Byreddy, A. R., Gupta, A., Barrow, C. J., & Puri, M. (2015). Comparison of cell disruption methods for improving lipid extraction from thraustochytrid strains. *Marine Drugs, 13*(8).
Carullo, D., Abera, B. D., Casazza, A. A., Donsì, F., Perego, P., Ferrari, G., & Pataro, G. (2018). Effect of pulsed electric fields and high pressure homogenization on the aqueous extraction of intracellular compounds from the microalgae *Chlorella vulgaris. Algal Research, 31*, 60–69.
Chen, C. L., Chang, J. S., & Lee, D. J. (2015). Dewatering and drying methods for microalgae. *Drying Technology, 33*, 443–454.

Chen, C. Y., Yeh, K. L., Aisyah, R., Lee, D. J., & Chang, J. S. (2011). Cultivation, photobioreactor design and harvesting of microalgae for biodiesel production: a critical review. *Bioresource Technology, 102*, 71–81.

Cheng, J., Qiu, Y., Zhang, J., Huang, R., Yang, W., & Fan, Z. (2017). Conversion of lipids from wet microalgae into biodiesel using sulfonated graphene oxide catalysts. *Bioresource Technology, 244*, 569–574.

Chew, K. W., Ling, T. C., & Show, P. L. (2019). Recent developments and applications of three-phase partitioning for the recovery of proteins. *Separation & Purification Reviews, 48*(1), 52–64.

Chintagunta, A. D., Kumar, A., Jeevan Kumar, S. P., & Verma, M. L. (2021). Contribution of metallic nanomaterials in algal biofuel production. In S. Rajendran, J. Qin, F. Gracia, & E. Lichtfouse (Eds.), *Metal and metal oxides for energy and electronics* (pp. 331–353). Springer International Publishing.

Dai, Y., Van Spronsen, J., Witkamp, G. J., Verpoorte, R., & Choi, Y. H. (2013). Ionic liquids and deep eutectic solvents in natural products research: Mixtures of solids as extraction solvents. *Journal of natural products, 76*(11), 2162–2173.

De Carvalho, J. C., Magalhães, A. I., de Melo Pereira, G. V., Medeiros, A. B. P., Sydney, E. B., Rodrigues, C., Aulestia, D. T. M., de Souza Vandenberghe, L. P., Soccol, V. T., & Soccol, C. R. (2020). Microalgal biomass pretreatment for integrated processing into biofuels, food, and feed. *Bioresource Technology, 300*, 122719.

De Jesus, S. S., Ferreira, G. F., Moreira, L. S., Wolf Maciel, M. R., & Maciel Filho, R. (2019). Comparison of several methods for effective lipid extraction from wet microalgae using green solvents. *Renewable Energy, 143*, 130–141.

de Souza Leite, L., dos Santos, P. R., & Daniel, L. A. (2020). Microalgae harvesting from wastewater by pH modulation and flotation: Assessing and optimizing operational parameters. *Journal of Environmental Management, 254*, 109825.

Dos Santos, R. R., Moreira, D. M., Kunigami, C. N., Aranda, D. A. G., & Teixeira, C. M. L. L. (2015). Comparison between several methods of total lipid extraction from *Chlorella vulgaris* biomass. *Ultrasonics Sonochemistry, 22*, 95–99.

Egesa, D., Chuck, C. J., & Plucinski, P. (2018). Multifunctional role of magnetic nanoparticles in efficient microalgae separation and catalytic hydrothermal liquefaction. *ACS Sustainable Chemistry & Engineering, 6*(1), 991–999.

Farid, M. S., Shariati, A., Badakhshan, A., & Anvaripour, B. (2013). Using nano-chitosan for harvesting microalga *Nannochloropsis* sp. *Bioresource Technology, 131*, 555–559.

Fattah, I. M. R., Noraini, M. Y., Mofijur, M., Silitonga, A. S., Badruddin, I. A., Khan,T. M. Y., Ong, H. C., & Mahlia, T. M. I. (2020). Lipid extraction maximization and enzymatic synthesis of biodiesel from microalgae. *Applied Sciences, 10*(17).

Fazril, I., Shamsuddin, A. H., Nomanbhay, S., Kusomo, F., Hanif, M., Ahmad Zamri, M. F. M., Akhiar, A., & Ismail, M. F. (2020). Microwave-assisted in situ transesterification of wet microalgae for the production of biodiesel: Progress review. *IOP Conference Series: Earth and Environmental Science, 476*, 12078.

Fu, Y., Hu, F., Li, H., Cui, L., Qian, G., Zhang, D., & Xu, Y. (2021). Application and mechanisms of microalgae harvesting by magnetic nanoparticles (MNPs). *Separation and Purification Technology, 265*, 118519.

Gao, S. S., Yang, J. X., Tian, J. Y., Ma, F., Tu, G., & Du, M. A. (2010). Electro-coagulation-flotation process for algae removal. *Journal of Hazardous Materials, 177*, 336–343.

Ge, S., Agbakpe, M., Zhang, W., & Kuang, L. (2015). Heteroaggregation between PEI-coated magnetic nanoparticles and algae: Effect of particle size on algal harvesting efficiency. *ACS Applied Materials & Interfaces, 7*(11), 6102–6108.

Geada, P., Rodrigues, R., Loureiro, L., Pereira, R., Fernandes, B., Teixeira, J. A., Vasconcelos, V., & Vicente, A. A. (2018). Electrotechnologies applied to microalgal biotechnology—Applications, techniques and future trends. *Renewable and Sustainable Energy Reviews, 94*, 656–668.

Gerardo, M. L., Hende, S. V. D., Han, V., Coward, T., & Skill, S. C. (2015). Harvesting of microalgae within a biorefinery approach: a review of the developments and case studies from pilot-plants. *Algal Research, 11*, 248–262.

Goh, B. H. H., Ong, H. C., Cheah, M. Y., Chen, W.-H., Yu, K. L., & Mahlia, T. M. I. (2019). Sustainability of direct biodiesel synthesis from microalgae biomass: A critical review. *Renewable and Sustainable Energy Reviews, 107*, 59–74.

Guldhe, A., Moura, C. V. R., Singh, P., Rawat, I., Moura, E. M., Sharma, Y., & Bux, F. (2017). Conversion of microalgal lipids to biodiesel using chromium-aluminum mixed oxide as a heterogeneous solid acid catalyst. *Renewable Energy, 105*, 175–182.

Guo, S. L., Zhao, X. Q., Wan, C., Huang, Z. Y., Yang, Y. L., Alam, M. A., et al. (2013). Characterization of flocculating agent from the self-flocculating microalga *Scenedesmus obliquus* AS-6-1 for efficient biomass harvest. *Bioresource Technology, 145*, 285.

Halim, R., Gladman, B., Danquah, M. K., & Webley, P. A. (2011). Oil extraction from microalgae for biodiesel production. *Bioresource Technology, 102*(1), 178–185.

Halim, R., Harun, R., Danquah, M. K., & Webley, P. A. (2012). Microalgal cell disruption for biofuel development. *Applied Energy, 91*(1), 116–121.

Hansel, P. A., Riefler, R. G., & Stuart, B. J. (2014). Efficient flocculation of microalgae for biomass production using cationic starch. *Algal Research, 5*, 133–139.

Harris, J., Viner, K., Champagne, P., & Jessop, P. G. (2018). Advances in microalgal lipid extraction for biofuel production: A review. *Biofuels, Bioproducts and Biorefining, 12*(6), 1118–1135.

He, M., Yan, Y., Pei, F., Wu, M., Gebreluel, T., Zou, S., & Wang, C. (2017). Improvement on lipid production by *Scenedesmus obliquus* triggered by low dose exposure to nanoparticles. *Scientific Reports, 7*(1), 1–12.

Hena, S., Fatihah, N., Tabassum, S., Lalung, J., & Jing, S. Y. (2016). Magnetophoretic harvesting of freshwater microalgae using polypyrrole/Fe_3O_4 nanocomposite and its reusability. *Journal of Applied Phycology, 28*(3), 1597–1609.

Ho, S.-H., Chiu, S.-Y., Kao, C.-Y., Chen, T.-Y., Chang, Y.-B., Chang, J.-S., et al. (2017). Ferrofluidassisted rapid and directional harvesting of marine microalgal *Chlorella* sp. used for biodiesel production. *Bioresource Technology, 244*, 1337–1340

Hu, Q., Pan, B., Xu, J., Sheng, J., & Shi, Y. (2007). Effects of supercritical carbon dioxide extraction conditions on yields and antioxidant activity of *Chlorella pyrenoidosa* extracts. *Journal of Food Engineering, 80*(4), 997–1001.

Huang, W.-C., & Kim, J.-D. (2016). Nickel oxide nanoparticle-based method for simultaneous harvesting and disruption of microalgal cells. *Bioresource Technology, 218*.

Hurtado, D. X., Garzón-Castro, C. L., Cortés-Romero, J., & Camacho, E. T. (2018). Comparison of lipid extraction methods for the microalgae *Acutodesmus obliquus*. *International Journal of Agricultural and Biological Engineering, 11*(5), 211–217.

Japar, A. S., Takriff, M. S., & Yasin, N. H. M. (2017). Harvesting microalgal biomass and lipid extraction for potential biofuel production: A review. *Journal of Environmental Chemical Engineering, 5*(1), 555–563.

Jeevan Kumar, S. P., Vijay Kumar, G., Dash, A., Scholz, P., & Banerjee, R. (2017). Sustainable green solvents and techniques for lipid extraction from microalgae: A review. *Algal Research, 21*, 138–147.

Kandasamy, G., & Shaleh, S. R. M. (2017). Harvesting of the microalga *Nannochloropsis* sp. by bioflocculation with mung bean protein extract. *Applied Biochemistry and Biotechnology, 182*, 1–12.

Kapoore, R. V, Butler, T. O., Pandhal, J., & Vaidyanathan, S. (2018). Microwave-assisted extraction for microalgae: From biofuels to biorefinery. *Biology, 7*(1).

Khanra, A., Vasistha, S., Kumar, P., & Rai, M. P. (2020). Role of C/N ratio on microalgae growth in mixotrophy and incorporation of titanium nanoparticles for cell flocculation and lipid enhancement in economical biodiesel application. *3 Biotech, 10*(8).

Kim, B., Praveenkumar, R., Lee, J., Nam, B., Kim, D.-M., Lee, K., Lee, Y.-C., & Oh, Y.-K. (2016). Magnesium aminoclay enhances lipid production of mixotrophic *Chlorella* sp. KR-1 while reducing bacterial populations. *Bioresource Technology, 219*, 608–613.

Kim, D.-Y., Lee, K., Lee, J., Lee, Y.-H., Han, J.-I., Park, J.-Y., & Oh, Y.-K. (2017). Acidified-flocculation process for harvesting of microalgae: Coagulant reutilization and metal-free-microalgae recovery. *Bioresource Technology, 239*, 190–196.

Koley, S., Prasad, S., Bagchi, S. K., & Mallick, N. (2017). Development of a harvesting technique for large-scale microalgal harvesting for biodiesel production. *RSC Advances, 7*, 7227–7237.

Kotasthane, T. (2017). Potential of microalgae for sustainable biofuel production. *Journal of Marine Science: Research & Development, 07*(01).

Krishnan, S., Ghani, N. A., Aminuddin, N. F., Quraishi, K. S., Azman, N. S., Cravotto, G., & Leveque, J.-M. (2020). Microwave-assisted lipid extraction from Chlorella vulgaris in water with 0.5%–2.5% of imidazolium based ionic liquid as additive. *Renewable Energy, 149*, 244–252.

Kumar, V., Arora, N., Nanda, M., & Pruthi, V. (2019). Different cell disruption and lipid extraction methods from microalgae for biodiesel production. In M. A. Alam & Z. Wang (Eds.) *Microalgae biotechnology for development of biofuel and wastewater treatment* (pp. 265–292). Springer Singapore.

Laamanen, C. A., Ross, G. M., & Scott, J. A. (2016). Flotation harvesting of microalgae. *Renewable and Sustainable Energy Reviews, 58*, 75–86.

Lai, Y. S., McCaw, A., Ontiveros-Valencia, A., Shi, Y., Parameswaran, P. & Rittmann, B.E. (2016). Multiple synergistic benefits of selective fermentation of Scenedesmus biomass for fuel recovery via wet-biomass extraction. *Algal Research, 17*, 253–260.

Lee, S. Y., Khoiroh, I., Vo, D.-V. N., Senthil Kumar, P., & Show, P. L. (2021). Techniques of lipid extraction from microalgae for biofuel production: A review. *Environmental Chemistry Letters, 19*(1), 231–251.

Lee, C., Chong, M., Robinson, J., & Binner, E. (2014). A review on development and application of plant-based bioflocculants and grafted bioflocculants. *Industrial and Engineering Chemistry Research, 53*, 18357–18369.

Lemos, J. S., Vargas, J. V. C., Mariano, A. B., Kava, V., & Ordonez, J. C. (2016). A flocculation strategy for harvesting high lipid content microalgae biomass. *IEEE Conference on Technologies for Sustainability (SusTech), 2016*, 240–245.

Li, S., Hu, T., Xu, Y., Wang, J., Chu, R., Yin, Z., Mo, F., & Zhu, L. (2020). A review on flocculation as an efficient method to harvest energy microalgae: Mechanisms, performances, influencing factors and perspectives. *Renewable and Sustainable Energy Reviews, 131*, 110005.

Liang, D., Alam, M. A., Lu, L., Fan, R., Xu, J., & Wu, J. (2022). Water-plasma-enhanced and phase-separation-assisted extraction of microalgal lipid for biodiesel production. *Bioresource Technology, 354*, 127198.

Liu, J. Y., Zeng, L. H., Ren, Z. H., Du, T. M., & Liu, X. (2020a). Rapid in situ measurements of algal cell concentrations using an artificial neural network and single-excitation fluorescence spectrometry. *Algal Research, 45*, 101739.

Liu, P., Wang, T., Yang, Z., Hong, Y., Xie, X., & Hou, Y. (2020b). Effects of Fe_3O_4 nanoparticle fabrication and surface modification on *Chlorella* sp. harvesting efficiency. *Science of the Total Environment, 704*, 135286.

Lopez-Exposito, P., Campano, C., van de Ven, T. G. M., Negro, C., & Blanco, A. (2019). Microalgae harvesting with the novel flocculant hairy cationic nanocrystalline cellulose. *Colloids and Surfaces B, 178*, 329–336.

Luo, S., Wu, X., Jiang, H., Yu, M., Liu, Y., Min, A., et al. (2019). Edible fungi-assisted harvesting system for efficient microalgae bio-flocculation. *Bioresource Technology, 282*, 325–330.

Mallick, N., Bagchi, S. K., Koley, S., & Singh, A. K. (2016). Progress and challenges in microalgal biodiesel production. *Frontiers in Microbiology, 7*(JUN), 1–11.

Mathimani, T., & Mallick, N. (2018). A comprehensive review on harvesting of microalgae for biodiesel—Key challenges and future directions. *Renewable and Sustainable Energy Reviews, 91*, 1103–1120.

Mehariya, S., Sharma, N., Iovine, A., Casella, P., Marino, T., Larocca, V., Molino, A., & Musmarra, D. (2020). An integrated strategy for nutraceuticals from *Haematoccus pluvialis*: From cultivation to extraction. *Antioxidants, 9*(9), 825.
Menegazzo, M. L., & Fonseca, G. G. (2019). Biomass recovery and lipid extraction processes for microalgae biofuels production: A review. *Renewable and Sustainable Energy Reviews, 107*, 87–107.
Mubarak, M., Shaija, A., & Suchithra, T. V. (2015). A review on the extraction of lipid from microalgae for biodiesel production. *Algal Research, 7*, 117–123.
Muhammad, G., Alam, M. A., Mofijur, M., Jahirul, M. I., Lv, Y., Xiong, W., Ong, H. C., & Xu, J. (2021). Modern developmental aspects in the field of economical harvesting and biodiesel production from microalgae biomass. *Renewable and Sustainable Energy Reviews, 135*, 110209.
Muradov, N., Taha, M., Miranda, A. F., Wrede, D., Kadali, K., Gujar, A., et al. (2015). Fungalassisted algal flocculation: Application in wastewater treatment and biofuel production. *Biotechnology for Biofuels, 8*, 24.
Nguyen TDP, Le TVA, Show PL, Nguyen TT, Tran MH, Tran TNT, et al. (2019). Bioflocculation formation of microalgae-bacteria in enhancing microalgae harvesting and nutrient removal from wastewater effluent. *Bioresource Technology, 272*, 34–39.
Nobre, B. P., Villalobos, F., Barragán, B. E., Oliveira, A. C., Batista, A. P., Marques, P. A. S. S., Mendes, R. L., Sovová, H., Palavra, A. F., & Gouveia, L. (2013). A biorefinery from *Nannochloropsis* sp. microalga—Extraction of oils and pigments. Production of biohydrogen from the leftover biomass. *Bioresource Technology, 135*, 128–136.
Noor, N., Enche Baharuddin, D., Shuhada, N., Naemah Sohif, H., Azlina, W., Karim, A., et al. (2016). Marine microalgae flocculation using plant: the case of *Nannochloropsis oculata* and *Moringa oleifera*. *Pakistan Journal of Botany, 48*, 831–840.
Pádrová, K., Lukavský, J., Nedbalová, L., Čejková, A., Cajthaml, T., Sigler, K., Vítová, M., & Řezanka, T. (2015). Trace concentrations of iron nanoparticles cause overproduction of biomass and lipids during cultivation of cyanobacteria and microalgae. *Journal of Applied Phycology, 27*(4), 1443–1451.
Panadare, D. C., & Rathod, V. K. (2017). Three phase partitioning for extraction of oil: A review. *Trends in Food Science & Technology, 68*, 145–151.
Pandey, A., Pathak, V. V., Kothari, R., Black, P. N., & Tyagi, V. V. (2019). Experimental studies on zeta potential of flocculants for harvesting of algae. *Journal of Environmental Management, 231*, 562–569.
Patel, A., Mikes, F., & Matsakas, L. (2018). An overview of current pretreatment methods used to improve lipid extraction from oleaginous microorganisms. *Molecules, 23*(7).
Patel, A., Matsakas, L., Sartaj, K., & Chandra, R. (2020). Extraction of lipids from algae using supercritical carbon dioxide. In Inamuddin, A. M. Asiri, & A. M. Isloor (Eds.), *Green sustainable process for chemical and environmental engineering and science* (pp. 17–39). Elsevier.
Paudel, A., Jessop, M. J., Stubbins, S. H., Champagne, P., & Jessop, P. G. (2015). Extraction of lipids from microalgae using CO_2-expanded methanol and liquid CO_2. *Bioresource Technology, 184*, 286–290.
Prommuak, C., Sereewatthanawut, I., Pavasant, P., Quitain, A. T., Goto, M., & Shotipruk, A. (2016). The effect of pulsed microwave power on transesterification of *Chlorella* sp. for biodiesel production. *Chemical Engineering Communications, 203*(5), 575–580.
Pugazhendhi, A., Shobana, S., Bakonyi, P., Nemestóthy, N., Xia, A., Banu, J. R., & Kumar, G. (2019). A review on chemical mechanism of microalgae flocculation via polymers. *Biotechnology Reports, 21*, e00302.
Rahul, R., Kumar, S., Jha, U., & Sen, G. (2015). Cationic inulin: a plant based natural biopolymer for algal biomass harvesting. *International Journal of Biological Macromolecules, 72*, 868–874.
Ren, H.-Y., Dai, Y.-Q., Kong, F., Xing, D., Zhao, L., Ren, N.-Q., Ma, J., & Liu, B.-F. (2020). Enhanced microalgal growth and lipid accumulation by addition of different nanoparticles under xenon lamp illumination. *Bioresource Technology, 297*, 122409.

Romero, N., Visentini, F. F., Márquez, V. E., Santiago, L. G., Castro, G. R., & Gagneten, A. M. (2020). Physiological and morphological responses of green microalgae *Chlorella vulgaris* to silver nanoparticles. *Environmental Research, 189*, 109857.

Salim, S., Bosma, R., Vermuë, M. H., & Wijffels, R. H. (2011). Harvesting of microalgae by bioflocculation. *Journal of applied phycology, 23*, 849.

Salim, S., Vermuë, M. H., & Wijffels, R. H. (2012). Ratio between autoflocculating and target microalgae affects the energy-efficient harvesting by bio-flocculation. *Bioresource Technology, 118*, 49.

Santana, A., Jesus, S., Larrayoz, M. A., & Filho, R. M. (2012). Supercritical carbon dioxide extraction of algal lipids for the biodiesel production. *Procedia Engineering, 42*, 1755–1761.

Sati, H., Mitra, M., Mishra, S., & Baredar, P. (2019). Microalgal lipid extraction strategies for biodiesel production: A review. *Algal Research, 38*, 101413.

Sheng, J., Vannela, R., & Rittmann, B. E. (2011). Evaluation of methods to extract and quantify lipids from Synechocystis PCC 6803. *Bioresource Technology, 102*(2), 1697–1703.

Shin, H.-Y., Shim, S.-H., Ryu, Y.-J., Yang, J.-H., Lim, S.-M., & Lee, C.-G. (2018). Lipid Extraction from *Tetraselmis* sp. microalgae for biodiesel production using hexane-based solvent mixtures. *Biotechnology and Bioprocess Engineering, 23*(1), 16–22.

Shin, H. Y., Ryu, J. H., Bae, S. Y., Crofcheck, C., & Crocker, M. (2014). Lipid extraction from *Scenedesmus* sp. microalgae for biodiesel production using hot compressed hexane. *Fuel, 130*, 66–69.

Singh, J., Jain, D., Agarwal, P., & Singh, R. (2020). Auxin and cytokinin synergism augmenting biomass and lipid production in microalgae *Desmodesmus* sp. JS07. *Process Biochemistry, 95*, 223–234.

Singh, G., & Patidar, S. K. (2018). Microalgae harvesting techniques: A review. *Journal of Environmental Management, 217*, 499–508.

Srinuanpan, S., Chawpraknoi, A., Chantarit, S., Cheirsilp, B., & Prasertsan, P. (2018). A rapid method for harvesting and immobilization of oleaginous microalgae using pelletforming filamentous fungi and the application in phytoremediation of secondary effluent. *International Journal of Phytoremediation, 20*, 1017–1024.

Steriti, A., Rossi, R., Concas, A., & Cao, G. (2014). A novel cell disruption technique to enhance lipid extraction from microalgae. *Bioresource Technology, 164*, 70–77.

Sun, Y., Huang, Y., Liao, Q., Xia, A., Fu, Q., Zhu, X., & Fu, J. (2018). Boosting *Nannochloropsis oculata* growth and lipid accumulation in a lab-scale open raceway pond characterized by improved light distributions employing built-in planar waveguide modules. *Bioresource Technology, 249*, 880–889.

Suparmaniam, U., Lam, M. K., Uemura, Y., Lim, J. W., Lee, K. T., & Shuit, S. H. (2019). Insights into the microalgae cultivation technology and harvesting process for biofuel production: A review. *Renewable and Sustainable Energy Reviews, 115*, 109361.

Taher, H., Giwa, A., Abusabiekeh, H., & Al-Zuhair, S. (2020). Biodiesel production from *Nannochloropsis gaditana* using supercritical CO_2 for lipid extraction and immobilized lipase transesterification: Economic and environmental impact assessments. *Fuel Processing Technology, 198*, 106249.

Tan, X. B., Lam, M. K., Uemura, Y., Lim, J. W., Wong, C. Y., & Lee, K. T. (2018). Cultivation of microalgae for biodiesel production: A review on upstream and downstream processing. *Chinese Journal of Chemical Engineering, 26*(1), 17–30.

Tan, Z. J., Yang, Z. Z., Yi, Y. J., Wang, H. Y., Zhou, W. L., Li, & F. F. (2015). Extraction of oil from flaxseed (Linum usitatissimum L.) using enzymeassisted three-phase partitioning. *Applied Biochemistry and Biotechnology*.

Tang, S., Qin, C., Wang, H., Li, S., & Tian, S. (2011). Study on supercritical extraction of lipids and enrichment of DHA from oil-rich microalgae. *The Journal of Supercritical Fluids, 57*(1), 44–49.

To, T. Q., Procter, K., Simmons, B. A., Subashchandrabose, S., & Atkin, R. (2018). Low cost ionic liquid–water mixtures for effective extraction of carbohydrate and lipid from algae. *Faraday Discussions, 206*, 93–112.

Ummalyma, S. B., Gnansounou, E., Sukumaran, R. K., Sindhu, R., Pandey, A., & Sahoo, D. (2017). Bioflocculation: An alternative strategy for harvesting of microalgae—An overview. *Bioresource Technology, 242*, 227.

Vargas-Estrada, L., Torres-Arellano, S., Longoria, A., Arias, D. M., Okoye, P. U., & Sebastian, P. J. (2020). Role of nanoparticles on microalgal cultivation: A review. *Fuel, 280*, 118598.

Vasistha, S., Khanra, A., & Rai, M. P. (2021). Influence of microalgae-ZnO nanoparticle association on sewage wastewater towards efficient nutrient removal and improved biodiesel application: An integrated approach. *Journal of Water Process Engineering, 39*, 101711.

Vasistha, S, Khanra, A., Clifford, M., & Rai, M. P. (2021). Current advances in microalgae harvesting and lipid extraction processes for improved biodiesel production: A review. Renewable and Sustainable Energy Reviews, *137*, 110498.

Vu, C. H. T., Chun, S.-J., Seo, S.-H., Cui, Y., Ahn, C.-Y., & Oh, H.-M. (2019). Bacterial community enhances flocculation efficiency of *Ettlia* sp. by altering extracellular polymeric substances profile. *Bioresource Technology, 281*, 56–65.

Wan, C., Alam, M. A., Zhao, X. Q., Zhang, X. Y., Guo, S. L., Ho, S. H., et al. (2015). Current progress and future prospect of microalgal biomass harvest using various flocculation technologies. *Bioresource Technology, 184*, 251–257.

Wang, Y., Guo, W., Yen, H., Ho, S., Lo, Y., Cheng, C., Ren, N., & Chang, J. (2015). Cultivation of *Chlorella vulgaris* JSC-6 with swine wastewater for simultaneous nutrient/COD removal and carbohydrate production. *Bioresource Technology, 198*, 619–625.

Wang, Q., Oshita, K., & Takaoka, M. (2021a). Effective lipid extraction from undewatered microalgae liquid using subcritical dimethyl ether. *Biotechnology for Biofuels, 14*(1), 17.

Wang, T., Fu, X.Z., Wang, S. (2021b). Etching oxide overlayers of NiFe phosphide to facilitate surface reconstruction for oxygen evolution reaction. *Green Energy & Environment, 3*, 500.

Wang, Y., Zhuo, S., Li, N., & Yang, Y. (2011). Influences of various aluminum coagulants on algae floc structure, strength and flotation effect. *Procedia Environmental Sciences, 8*, 75–80.

Wu, Y.-H., Shen, L.-C., Hu, H.-Y., Hankins, N. P., & Huang, W. E. (2018). An efficient microalgal biomass harvesting method with a high concentration ratio using the polymersurfactant aggregates process. *Algal Research, 30*, 8.

Yoo, C., La, H. J., Kim, S. C., & Oh, H. M. (2015). Simple processes for optimized growth and harvest of *Ettlia* sp. by pH control using CO_2 and light irradiation. *Biotechnology and Bioengineering, 112*, 288–296.

Zhang, H., & Zhang, X. (2019). Microalgal harvesting using foam flotation: A critical review. *Biomass and Bioenergy, 120*, 176–188.

Zhang, Y., Soldatov, S., Papachristou, I., Nazarova, N., Link, G., Frey, W., & Silve, A. (2022). Pulsed microwave pretreatment of fresh microalgae for enhanced lipid extraction. *Energy, 248*, 123555.

Zhou, W., Wang, Z., Alam, M. A., Xu, J., Zhu, S., Yuan, Z., Huo, S., Guo, Y., Qin, L., & Ma, L. (2019). Repeated utilization of ionic liquid to extract lipid from algal biomass. *International Journal of Polymer Science*, 9209210.

Zhou, W., Ruan, R., & Wang, J. (2015). Bio-flocculation of microalgae: Status and prospects. *Current Biotechnology, 4*(4), 448–456.

Zhu, L., Li, Z., & Hiltunen, E. (2018). Microalgae *Chlorella vulgaris* biomass harvesting by natural flocculant: Effects on biomass sedimentation, spent medium recycling and lipid extraction. *Biotechnol Biofuels, 11*, 183.

Zou, X., Li, Y., Xu, K., Wen, H., Shen, Z., & Ren, X. (2018). Microalgae harvesting by buoy-bead flotation process using bioflocculant as alternative to chemical Flocculant. *Algal Research, 32*, 233–240.

Chapter 7
Microalgae Biomass Conversion Process

Depending on the type of biomass, the alternatives for conversion, and the intended use of the energy, there are various ways for turning algae into energy. There are two forms of microalgal biomass conversion (Fig. 7.1):

(i) thermal process;
(ii) biochemical process.

The quality of the biomass, the feedstock, and the desired energy content are influencing factors for conversion.

7.1 Thermochemical Conversion

Biomass is thermally broken down in this process to create fuel products. The process uses pyrolysis, gasification, liquefaction, and combustion.

7.1.1 Gasification

In this process, hydrocarbons are transformed into a combustible gaseous mixture (syngas), while oxygen is in short supply. It can be used as fuel for gas engines or for direct combustion. Its calorific value is only 4–6 MJ/m^3, though. Fuels can be transported through syngas using one of two alternative processes: either the water gas shift reaction, which produces hydrogen, or the Fischer TrOpen pond ch reaction, which transforms hydrocarbons into liquid fuels and chemicals like methanol. For gasifying microalgae biomass into biofuel, several studies have been published. In order to ascertain the gas composition necessary to produce methanol, Hirano et al. examined partially oxidized *Spirulina* at high temperatures between 850 and 1000 °C. The maximum theoretical yield of 1000 was 0.64 g of methanol for 1 g of biomass.

M. P. Rai and S. Vasistha, *Microalgae Biotechnology for Wastewater Treatment, Resource Recovery and Biofuels*, https://doi.org/10.1007/978-3-031-31674-6_7

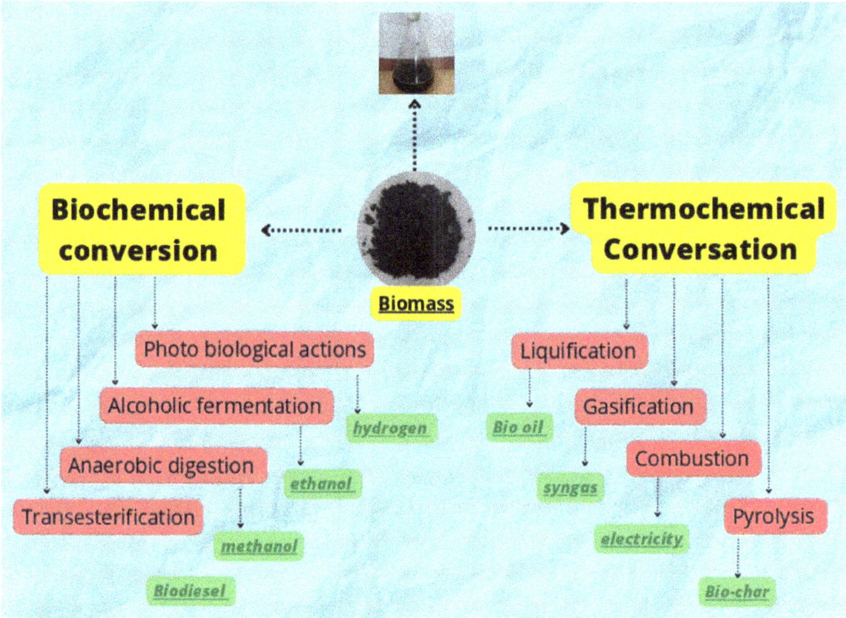

Fig. 7.1 Biomass conversion process

7.1.2 Liquefaction

Wet algal biomass is utilized in this technique to create liquid gasoline. Mechanically harvested microalgae cells have a high moisture content and are utilized as a starting point for liquefaction. This procedure typically results in high viscosity oils and calls for biomass, solvents, gases like H_2 or CO, as well as catalysts. Wet biomass can be consumed through liquefaction, which transforms it into smaller molecules with higher energy densities. *B. braunii* liquefied at 300 °C with a yield of 64% (dry weight). A 42% dry weight oil yield was produced by *Dunaliella tertiolecta.* This demonstrates unequivocally that liquefaction is a crucial alternative for biomass conversion.

7.1.3 Pyrolysis

Algal biomass can be used to create a variety of goods at 350–700 °C without air, including biochar, bio-oil, and syngas. Biomass can be pyrolyzed into liquid fuels and scaled up for large-scale production. One of the most crucial processes for replacing petroleum in future with liquid biofuels is flash pyrolysis, which takes place at 500 °C. In light of the fact that it can attain a high biomass to liquid ratio of about 95.5%. However, there are still some significant issues that must be resolved before microalgal pyrolysis may be commercialized. This technique produces oil that is acidic in nature and extremely viscous and contains both solids and water. In order to remove alkali and reduce the oxygen concentration, the procedure also needs advancements in hydrogenation and catalytic cracking.

Three stages are involved in traditional process.

Pre-pyrolysis is the name for the initial stage of biomass degradation at 550–950 K. In this process, the water is eliminated, the link is broken, and subsequently, carbonyl and carboxyl groups are created. In the second stage, solid breakdown leads to the formation of pyrolysis products. Char breaks down into carbon-rich residues in the third stage. At 850–1250 K, a quick pyrolysis process takes place, which causes biomass to breakdown and produce pyrolysis products. The product's composition is bio-oil (60–70%), biochar (15–25%), and non-condensed gases (10–20%).

7.1.4 Combustion

Combustion is one of the simplest ways to generate energy in all thermochemical processes. The fuel is often burned in an environment with too much air, which causes the fuel to react and cause burning. CO_2, H_2O, and heat release are the main byproducts. Typically, combustion takes place at temperatures of 800 °C or higher. The boilers and furnaces where the steam needed to make the electricity frequently experience it. Algal biomass combustion results in decreased CO_2 emissions, according to the life cycle assessment (LCA). But for the technology to be scaled up further, more research and a modified burner to deal with algae are needed.

7.2 Biochemical Conversion

Biofuels are created by converting biomass through biological processes such anaerobic digestion, fermentation, and transesterification.

7.2.1 Anaerobic Digestion (AD)

This process transforms organic material into gas and energy. Wet algal biomass is best suited for the AD process, which is one of the most effective methods for greater moisture feed. The three stages of this process are methanogenesis, fermentation, and hydrolysis. In the initial stage, complex molecules are converted into soluble sugars. After that comes fermentation, which turns soluble sugar into alcohols, acetic acid, and mostly hydrogen and carbon dioxide gas. During the third stage, methanogens convert these gases into methane and carbon dioxide. Experimental evidence provided by the considerable increase in methane production which was found following AD when combined with paper waste and algal biomass. By substituting 50% of waste paper for pure algal biomass, the rate of methane production was doubled. However, the products' heating value is reduced by 20–40%.

7.2.2 Fermentation

Alcohols like bioethanol are produced during fermentation from starch or cellulose. Yeast is typically used to turn carbohydrates into ethanol. For product separation and alcohol purification, distillation is necessary. Gasification can make use of the fermentation waste. Due to the process cost reduction, this procedure is now more commercially viable. Microalgae like *Chlorella Vulgaris* are the most common species to produce ethanol due to high starch content, and maximum ethanol conversion of 65% reported.

7.2.3 Photobiological Process

A safe and effective energy source is hydrogen. To create hydrogen gas, microalgae must possess specific enzyme properties. Under anaerobic conditions, water molecules are transformed into oxygen and hydrogen ions during the photosynthesis of microalgae. Instead of hydrogenase enzymes, which transform hydrogen ions into hydrogen gas, eukaryotic microalgae are employed in this process. In essence, two processes produce hydrogen through photosynthesis. In the initial step, the hydrogen synthesis process is divided into two stages:

(1) normal algal growth;
(2) anaerobic conditions that encourage continuous hydrogen production in microalgae. This procedure has the benefit of not producing any poisonous or damaging compounds. Additionally, it offers beneficial consequences due to biomass.

Through photosynthesis, the photobiological activity can also simultaneously produce hydrogen and oxygen. The photosynthetic H_2O oxidation electron is released using this technique. The hydrogenase-mediated H_2-evolution mechanism receives the electron directly. Due to simultaneous production, this procedure is more efficient than a two-step approach. The theoretical yield for producing hydrogen gas in a two-step method was 198 kg/h/day.

7.2.4 Photosynthetic Microbial Fuel Cells

Microbial fuel cells are bioelectrochemical systems that use the anaerobic biodegradation of organic material to produce power. It has been demonstrated that combining microalgae with photosynthetic microbial fuel cells can remove CO_2 from the atmosphere and create an environment that is rich in oxygen. Anode and cathode of the photosynthetic microbial fuel cell are separated by a proton exchange membrane. By oxidizing the organic compounds, the bacteria in the anode produce electrons, which are then transmitted to the cathode electrode via an external circuit to generate power. This technology has the advantage that biodegradable wastes can be treated by microorganisms in the anode.

Chapter 8
Biorefinery Concept and Value-Added Products

8.1 Biorefinery Concepts

Biomass has long been utilized in the production of food, animal feed, and other industrial goods. The idea of a biorefinery has just recently come into existence, offering flexibility in terms of the various uses and outputs of biomass. The term "biorefinery" has had many definitions over the years. The most typical is that a biorefinery is a facility that integrates biomass conversion processes and equipment to produce a variety of biofuels and bioproducts, according to NREL. According to some definitions, biorefineries are groups of connected businesses or processes that share expenses and work together synergistically to make the economic model more workable (i.e., industrial symbiosis). Although the economy is mostly dependent on the tiny amount but high-value products, there are always energy or fuel products involved because the concept is built on the extrapolation of the petroleum refinery. There are numerous types of biorefineries that have been described, including lignocellulosic biorefineries (based on lignocellulosic biomass), whole plant biorefineries (based, e.g., on whole corn) green biorefineries (based on grasses), and oily biorefineries (based, e.g., from oily or rapeseed crop).

Researchers have proposed the idea of microalgae biorefineries, which entail converting algal biomass into biofuels and value-added goods. Increasing the financial viability of microalgae is a key driver for algal biorefineries. Because microalgae cultivation depends on the marginal economy, adding variety of products with an emphasis on high value-added ones could boost economic viability. Algal biorefinery technologies, however, also result in improved resource recovery, process efficiency, and cost-effectiveness to create valuable bioproducts in a closed loop. Microalgae have also been investigated for their ability to clean up greenhouse emissions, such as cleaning NOx and SOx from flue gas or absorbing and recycling CO_2. As a result, the microalgal biomass can be recycled to provide a usable product in addition to the environmental benefit of CO_2 collection. The potential uses in the framework of

M. P. Rai and S. Vasistha, *Microalgae Biotechnology for Wastewater Treatment, Resource Recovery and Biofuels*, https://doi.org/10.1007/978-3-031-31674-6_8

biorefinery are determined by the composition of the microalgae biomass. Polysaccharides, lipids, proteins, pigments, and other phenolic chemicals are the most crucial microalgal constituents.

The technology for the biorefinery approach is planned in accordance with the desired products. For instance, high oil-containing algae will be chosen for biodiesel synthesis. Then, the de-oiled microalgal biomass generated after lipid extraction could be further valorized to various applications through direct utilization (biosorption, fertilizer, feed supplement), biochemical (anaerobic digestion: biogas; fermentation, bioethanol), and thermochemical (pyrolysis: bio-oil, biochar; hydrothermal liquefaction: biocrude; gasification: syngas) conversion route.

8.2 Value-Added Products

8.2.1 *Pigments*

Carotenoids, chlorophylls, and phycobiliproteins are the three main categories of natural pigments found in microalgae. These pigments have been used in food, animal feed, additives, cosmetics, pharmaceutical industries, food coloring agents, and biomaterials as precursors of vitamins. Fat-soluble pigments known as carotenoids are regarded as accessory pigments since they provide color to other plant components. Historically, organic solvents and Soxhlet extractions were used to extract carotenoids from microalgae where the choice of solvent is important to define the level of affinity and encourage the release of carotenoids. Supercritical fluid extraction (SFE) is an alternative, albeit low selectivity and high solvent requirements make it necessary. Da Silva et al., 2016 also extracted carotenoids using a co-solvent-modified SC-CO_2 technique. Co-solvents were added, and it was concluded that this improved the carotenoid extraction process. Additionally, it was advised to use an anti-solvent precipitation method to increase the amount of carotenoids in the particles.

Chlorophylls are the principal photosynthetic pigments found in a wide variety of vegetables and fruits. They are lipid-soluble pigments with low polarity. Chlorophyll extraction traditionally involves a number of processes using organic solvents, much as the extraction of carotenoids. The type of solvent used, the method for disrupting the cell wall, the amount of time spent extracting, and the application of various empirical correlations are some aspects in solvent extraction that may affect the yield of pigments.

In comparison with acetone, methanol and ethanol were shown to have a higher extraction efficiency. Because chlorophyll and carotenoids have low polarity, SC-CO_2 has also been recommended as a good solvent for extracting these substances. However, the SC-CO_2 extraction operation needs a temperature range of 50–60 °C and a high pressure range of 300–500 bar. The primary photosynthetic accessory pigments in microalgae are phycobiliproteins. Natural colors in the food industry and pharmaceutical products are two of the main uses for phycobiliproteins. It is a promising substance for use in health-related applications because of its antioxidant, antiviral, anticancer, anti-allergic, anti-inflammatory, and neuro-protecting characteristics. Centrifugation, drying, homogenization, and a series of consecutive freeze-thaw processes can be used to generate phycobiliproteins. In the nutraceutical, food, and feed supply industries, carotenoids like astaxanthin are powerful antioxidants because of their robust anti-aging, sun protection, anti-inflammatory, and immune system-boosting properties.

8.2.2 *Vitamins*

In comparison with commercial feeds, microalgae have higher concentrations of vital vitamins and trace elements that are also bioavailable. In order to assess the effects of these components when consumed by the body, bioavailability studies on these components are frequently more important than feed quality. Nitrogen (N) availability is necessary for the generation of vitamins from microalgae. It was discovered that in low-N settings as opposed to N-replete conditions, cyanobacteria produced less vitamin B12 per cell. The type and quantity of nitrogen present in the culture medium also had an impact on the vitamin content.

Riboflavin is a different vitamin that can be found in microalgae and is crucial for animals raised in aquariums. Both the stationary and logarithmic growth phases of several microalgae were used to assess their composition. The cultures were first filtered under vacuum, and then, riboflavin was extracted from the microalgae sample under dim lighting. All tested species' cultures had more riboflavin after the stationary phase was initiated.

8.2.3 *Polyunsaturated Fatty Acids*

Numerous heart problems can be prevented with the help of polyunsaturated fatty acids (PUFA), which are widely acknowledged as important nutritional components. The growing depletion of marine resources has made it difficult to recover these molecules from ocean fish, and the need for PUFAs has driven researchers to find other sources of EPA and DHA. PUFAs, which are important for human nutrition and health, can be produced by microalgae. Because microalgae are a sustainable source, can be grown year-round, have a high oil content, and produce a lot of biomass, advanced microalgae technology can be used to produce PUFAs.

To assess the fatty acids in marine microalga, a number of techniques have been used, including direct saponification, Bligh and Dryer extraction, solvent extraction and sonication, and supercritical fluid extraction (SFE). The Bligh and Dryer technique extracts using a chloroform-methanol solution. Ethanol serves as the solvent and a catalyst during direct saponification. The SC-CO_2 approach produced the highest fatty acid levels. A combination of LED-Blue and LED-Red lighting increased the productivity of EPA in photobioreactors with the best light source arrangements, which can also be employed to increase EPA output.

8.2.4 *Biodiesel Production from Microalgae*

Biodiesel is a substitute for diesel and is defined as alkyl ester of long fatty acid chain obtained from vegetable oil, animal fats, and compounds containing TAGs, through transesterification reaction in presence of acid or base catalyst (Fig. 8.1) (Karpagam et al., 2021).

Regulations and standards guidelines for biofuel have been stated in some countries like for Testing and Materials ASTM-D6751, EN-14214, and Brazilian ANP 45/214 (Karpagam et al., 2021).

Common microalgae explored for biodiesel production are cyanophyte, chlorophyceae (green algae), and diatomaceous groups. Species like *Chlorella, Dunaliella* sp., and *Scenedesmus* sp. observed have high lipid content of 20–30%, 17–67%, and 11–55%, respectively. Previous studies observed lipid content of 55 % *Chlorella protothecoides* under heterotrophic condition, whereas *Chlorella sorokiniana* showed 55% lipid content when glucose as substrate. After the lipid is extracted from microalgae, it is trans-esterified along with acid or base catalyst (Islam et al., 2013; Nascimento et al., 2013; Sydney et al., 2011).

Prospect on Biofuel Production

In the past few decades, researchers are focusing to solve the environmental issues including wastewater treatment, energy crises, and air pollution. On a side note, microalgae have emerged as the solutions to these problems. Literature stated that microalgae-based biofuel technology offers several advantages such as (Ananthi et al., 2021; Fernández et al., 2021)

(a) Microalgae offers high growth rate as compared to plants used for biofuel; hence, the biomass productivity is higher in microalgae.
(b) It was estimated that the oil produced from microalgae is 16–17 times higher than sunflower, palm which can be used in biodiesel application.

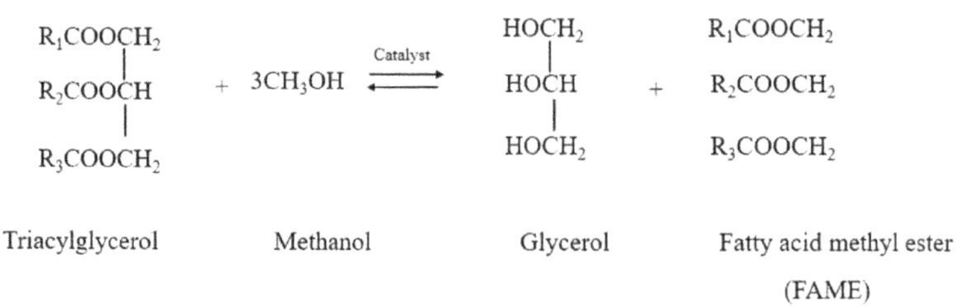

Fig. 8.1 Transesterification reaction

(c) Microalgae biomass is preferable over land-based crop for biodiesel production, and there is less competition for food and water.
(d) Microalgae can be used for the treatment of wastewater.
(e) They have the tendency to absorb carbon dioxide.

8.2.4.1 Catalyst

Catalyst is used to enhance the rate of reaction, and its selection is important for achieving high yield of biodiesel (Table 6.2). Depending on the fatty acid content in microalgae, the catalyst can be acidic or basic. In presence of a basic catalyst, the alcohol releases the nucleophilic alkoxide to attack the carbonyl part of TAGs which is electrophilic (Endalew et al., 2011a). On the contrary in acidic reaction, TAGs is protonated and alcohol will attack the carbon. Homogeneous catalyst or acid catalyst dissolves in the product and becomes difficult to separate (Sivaramakrishnan & Incharoensakdi, 2018). The acid catalyst gives low yield whereas base catalyst fastens the kinetics reaction and gives better yield. Some common catalysts used for transesterification are NaOH, KOH, CaO, and MgO (Veillette et al., 2017). Other catalysts like H_2SO_4 are preferred as they provide high FAME yield. The catalyst like CaO can be produced from egg shells, and chicken bone is attracting the researchers. It can be used for microalgae with low free fatty acid content as it is a base catalyst (Goh et al., 2019).

Chemical Homogeneous Catalyst

Various homogeneous catalysts like NaOH, HCL, H_2SO_4, and K_2HPO_4 are explored for biodiesel application (Galadima & Muraza, 2014). These catalysts are difficult to separate because of its homogeneity and add up the additional cost to the process. Studies reported that acid catalyst proved to be more beneficial than the alkaline catalyst (Skorupskaite et al., 2016). The higher FAME yield of 92% was observed using methanol-based transesterification with sodium hydroxide catalyst in *Chlorella* sp. (Velasquez-Orta et al., 2013). Whereas alkali catalysts such as KOH and NaOH increased the biodiesel yield from 0.5 to 2.5%.

It is worthy to note that NaOH has fewer free fatty acids as compared to KOH; hence, NaOH is preferred more in contrast to KOH (Ginting et al., 2012). In addition, NaOH is cheaper than KOH. (Hidalgo et al., 2015) reported that with increasing the concentration of catalyst, and the product yield is also increased.

Chemical Heterogeneous Catalyst

A heterogeneous catalyst enhances the in situ transesterification. A variety of catalysts includes CaO, MgO, Bao, and SrO for enhancing the reaction (Endalew et al., 2011b). In a study, the different concentration of ZrO_2/WO_3 upregulated the biodiesel

Table 8.1 Biodiesel synthesis from microalgae using various catalysts

Microalgae	Catalyst and reaction conditions	Biodiesel yield (%)	References
Euglena sanguinea	Calcium methoxide, temp; 49.85 °C; reaction time 1.5 h	94.83	Arunachalam Sivagurulingam et al. (2019)
Scenedesmus obliquus	Chromium-aluminum mixed oxide; temp: 80 °C; reaction time: 4 h	98.28	Guldhe et al. (2017)
Anabaena PCC 7120	Barium titanium oxide temp: 65 °C; reaction time: 3 h	98.41	Singh et al. (2019)
NannochlorOpen pond is gaditana	16 L of water, 36.4 L hexane, 36.4 L methanol, 18.2 L 98% sulfuric acid catalyst; reaction time: 2 h	85.5	Torres et al. (2017)
NannochlorOpen pond is oculata	Bi_2O_3/ZrO_2; temp: 80 °C; reaction time: 6 h	67.51	Rahman et al. (2020)
NannochlorOpen pond is oculata	Poly ethylene glycol (PEG) encapsulated $ZnOMn^{2+}$; temp: 60 °C; reaction time: 4 h	87.5	Raj et al. (2019)
Scenedesmus armatus	CaO nanocatalyst^{+}; temp: 75 °C; reaction time: 4 h	90.44	Pandit and Fulekar (2019)

yield. A recent report, CaO/Al_2O_3 in a ratio of 3:1, was used as a catalyst and produced the biodiesel yield of 89.9% with 98% purity (Kesserwan et al., 2020). In another study, CaO was used as catalysts immobilized onto polymeric mats for biodiesel production (Wilkanowicz et al., 2020). Heterogeneous catalysts can be easily separated from liquid products and perform similar functions as homogeneous catalysts. Another advantage of heterogeneous catalysts is that they have longer shelf life and can be reused. The most preferred catalysts are the group II metal oxides following the order as MgO, CaO, SrO, and BaO (Singh et al., 2020) (Table 8.1).

Economic Aspect of Microalgae Biofuel

According to the international energy outlook, the consumption of petroleum and other liquid fuel rises from 90 barrels/day in 2012 to 121 million barrels/day in 2040. India is the 4th largest energy consumer (Ansari et al., 2020). Approximately, 70% of the biodiesel is consumed alone by the transportation sector. Fuel demand has rapidly increased in recent times, and the US employs 50 gallons of biodiesel in a year (Yusoff et al., 2021). For the economic analysis of biofuel production from microalgae, we need to analyze the cost involved at every step. Majorly, the sub-processes included are:

- cultivation of microalgae in open and closed system;
- concentrating the microalgae biomass by various techniques;
- extraction of oils from microalgae biomass;
- conversion of oil from biomass to biofuel.

The major bottleneck in microalgae-based biodiesel production is the cost involved in the feedstock. The industry data revealed that the biofuel conversion plant for 5 million gallons per year would cost nearly $ 7,300,000 in 2008. According to the previous studies, the cost estimated for an open system for the productivity of 67.5 mt/ha/yr (Banu et al., 2020). The capital cost for the design was $ 39,850/ha, and the operational cost was estimated as $ 18,496/ha. In addition, the algal oil costs were estimated as $ 115/ha. The price of microalgae biofuel varies from 2.41$ to 6.09$ per gallon (Muhammad et al., 2021). Till now, the cost associated with microalgae biofuel is comparatively higher than petroleum fuel. Chia et al. (2018) studied the cost of 10,000 tons of microalgae biomass with lipid yield of 30% was $ 2.80/L; in contrast, the US price of fuel is $ 1.10/L.

In India, the biodiesel consumption is estimated to be 450 million barrels per year. Several agencies in India, such as IREDA and BDAI, are promoting the biodiesel production in terms of research and development (Sun et al., 2019). India's new biofuel policy aims to achieve 5% blending of biodiesel with conventional diesel by 2030. National Biodiesel Mission selected Jatropha as suitable oilseed for the blending to the conventional diesel by 2017 (Saravanan et al., 2018). However, Jatropha faced some economic issues and seasonal implications which restricted its utilization. Commercializing the algae-based biofuel also relies on the government subsidies and the future price of the oil (Su et al., 2017). Studies stated that the average price of crude oil prices increased from $ 26/barrel to $ 100/barrel from 2002 to 2008 at the annual rate of 25%, although the continuous elevated price of oil can affect the profitability of microalgae biodiesel (Raeisossadati et al., 2019) (Fig. 8.2).

8.2.5 Bioethanol

Bioethanol is made by digesting the sugars found in microalgae. Sugar concentration in microalgae varies based on algal strain and growing circumstances. Sadane-Bchir et al. isolated microalgae cultivated in slaughterhouse wastewater containing 80% to 85% sugars, extracted maximum sugar (62% total sugar and 32% reducing sugar) using a basic 0.75% NaOH solution, and converted to bioethanol 43% (w/w) based on the initial dry weight of biomass. Qu et al. (2020) demonstrated the cultivation of the microalga *Chlamydomonas* sp. QWY37 in swine wastewater with an 81% COD removal efficiency, a 96% TN removal efficiency, and a 100% TP removal efficiency, as well as the highest carbohydrate productivity (944 mg $L^{-1}d^{-1}$), which was then

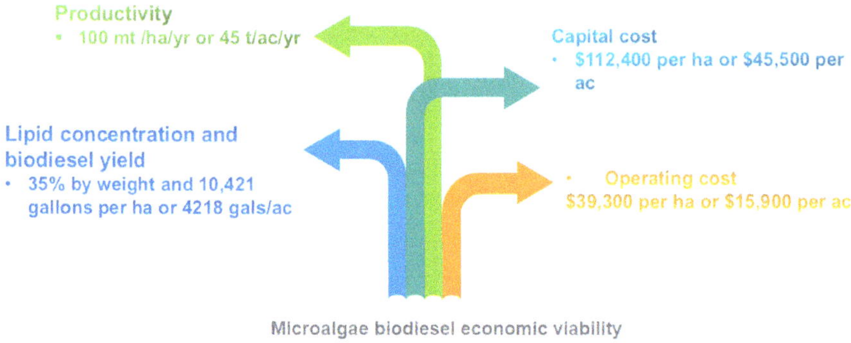

Fig. 8.2 Estimation of microalgae-based biofuel production

converted to bioethanol (61 g/L) via engineered yeast. However, because of the high production costs of microalgae, the typically limited levels of sugars in microalgae, and the complex process, which requires several costly unit operation steps, using microalgae for bioethanol synthesis is impracticable and unsustainable. Bioethanol synthesis from de-oiled biomass is a more appealing technique than single-product biorefinery designs. Bioethanol production from microalgae, on the other hand, is far from cost-effective. Tsolcha et al. (2021) use cyanobacterial biomass grown on agro-industrial wastewater (DWW, winery wastewater, Mix winery-raisin wastewater) in combination with raisin residue extract to produce bioethanol with a theoretical yield of 85.9%.

8.2.6 *Biohydrogen and Biogas*

Anaerobic digestion is a well-proven and reliable method for recovering bioenergy from biomass and wastewater. Organics are transformed to CO_2 and CH_4, the major components of biogas, during anaerobic digestion. Using anaerobic digestion on algal biomass with a lipid concentration less than 40% is more energy efficient. To recover energy from microalgae cultivated on wastewater resources, anaerobic digestion is often preferred over biodiesel synthesis.

Furthermore, anaerobic digestion can be used with biomass that has a high moisture content (80–90%), which lowers the cost of biomass dewatering compared to other biofuel applications.

Microalgae biomass typically has a C:N ratio of around 10 which is substantially lower than the normal anaerobic digestion substrate, which has a C:N ratio of 20 to 30. For mono-digestion, ammonia-tolerant cultures are recommended. Co-digestion of microalgae biomass with carbon-rich biomasses, such as activated sludge, is another option. Compared to the digestion of sludge and microalgae separately, Wágner et al. (2016) encouraged co-flocculation of green microalgae and activated sludge, which considerably reduced harvesting costs and increased biomethane conversion potential.

Microalgae can also be used to make biohydrogen. The synthesis of biohydrogen from microalgae can be done in two ways. Microalgae can manufacture hydrogen in one approach by autotrophically using light energy in their metabolism. *Chlamydomonas reinhardtii* has been widely studied and is regarded as one of the most promising H_2 producers. However, the productivity is considerably too low to offset the process costs, and so biohydrogen production by autotrophic growth is not commercially viable. The bacterial conversion of carbohydrates in microalgae to volatile fatty acids, CO_2, and H_2 at low pH is the second way. Dark fermentation uses solely the sugars in the microalgae to produce biohydrogen. Biohydrogen production, like bioethanol, would be prohibitively expensive; therefore, using microalgae for biohydrogen synthesis is unlikely to become a practical reality.

8.2.7 Biofertilizer

Microalgae cultivation is far less resource-intensive than physicochemical methods for producing fertilizer from wastewaters, such as struvite or calcium phosphate, because it does not involve the use of chemicals or energy-intensive procedures. Microalgae convert phosphate and nitrogen from wastewater into algal biomass. Microalgae act as a slow-release organic fertilizer that keeps nitrogen, phosphate, and potassium from being lost. Microalgae, on the other hand, have other advantages than providing nutrients to plants. Extracellular polymeric substances (EPS) are produced by microalgae and aid in the formation of biofilms in soil. Biofilms help to provide soil structure, retain water, and reduce erosion caused by over-farming. Polysaccharides are found in EPS and give organic carbon to the soil microbiota. Using algae as fertilizers has been shown in several studies to result in a more active rhizosphere. Some algal metabolites (e.g., indole acetic acid) help plants grow by stimulating root growth, nutrient mobilization, and relieving abiotic stress. Furthermore, microalgae store a high number of proteins, including tryptophan and arginine, which are necessary amino acids. Algae can boost the amount of chlorophyll and carotenoids in plants, which boosts photosynthetic activity and increases fruit quality. Finally, several algae secondary metabolites exhibit antibacterial activity, allowing chemical pesticides to be replaced. Umamaheswari and Shanthakumar (2021) used *C. pyrenoidosa* to remediate paddy-soaked rice mill effluent. The maximum elimination of ammoniacal nitrogen and phosphate was determined to be 70% and 65%, respectively. The biomass produced comprised a lipid content of 35%, a protein content of 30%, and a carbohydrate content of 9%. They observed that using this live algal biomass as a soil-adjusted biofertilizer on Indian okra (Abelmoschus angulosus) resulted in plants with improved chlorophyll, fresh and dry shoot, and root weights that were 30% better than soil amended with chemical fertilizers (Fig. 8.3).

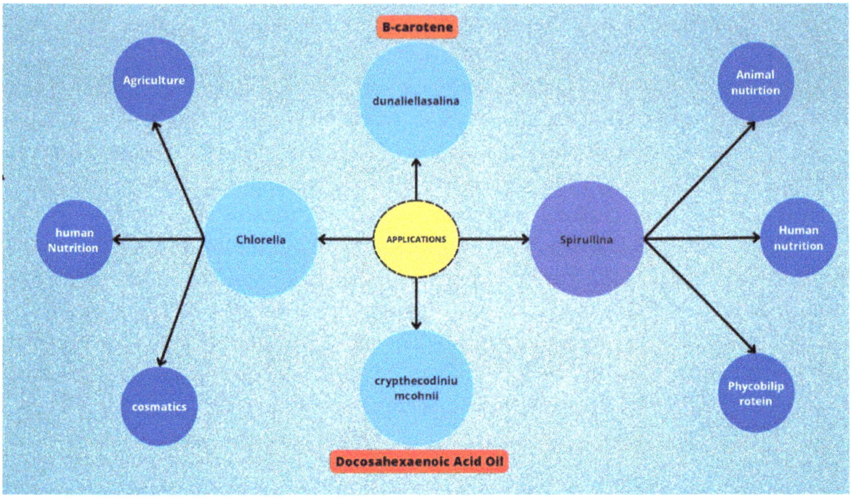

Fig. 8.3 Various applications of microalgae in various fields

8.2.8 Food and Feed

Microalgal biomass is high in amino acids, polyunsaturated fatty acids, and antioxidant pigments, all of which are high in nutritional value and have positive effects on human and animal health. Microalgae biomass has long been a part of human diets and has been seen as an alternate food and feed source, as well as a possible solution to food scarcity. Over 75% of the grown microalgae biomass is now used as a human food supplement in the form of powder, pills, and capsules. To be used as food and feed, the microalgae culture is required to be free from micropollutants such as toxic chemicals, heavy metals, and pathogenic bacteria and viruses. Thus, the pretreatment of IWW and closed cultivation systems are required for feed and food microalgae production, leading to high construction and operational cost.

One option for dealing with the problem is to employ "clean" industrial wastewaters (IWW) that are free of heavy metals, organic pollutants, and microorganisms. Rasouli et al. described a new method for recovering nutrients from potato processing IWW by using it as a growing medium for *Chlorella sorokiniana*, which contains up to 52.5% DW of feed-grade microbial protein. Returning the microalgae biomass to the IWW source is an alternative solution. For example, the RAS mentioned in the previous section might feed shrimp, oysters, and lower trophic fish aquaculture wastewater-grown microalgae.

8.2.9 Nutraceutical and Pharmaceutical

To obtain the bioactive chemicals for use in healthcare goods for both humans and animals, some microalgae species are industrially grown. Chlorophyll, carotenoids, astaxanthins, zeaxanthins, and phycobilins are examples of pigments that are promising antioxidants that have the potential to increase immune function as well as have potent anti-inflammatory and anti-aging properties. They can be used in the dietary supplements, food, and animal feed industries. Additional sources of bioactive chemicals that can be used as natural medicinal agents include microalgal polysaccharides and protein hydrolysates. Due to the presence of long-chain polyunsaturated fatty acids in the microalgal lipid compositions, interest in them is also growing (LC-PUFAs). LC-PUFAs are important nutrients with immunological functions and a variety of uses in pharmaceuticals and nutraceuticals for the cosmetics industry. LC-PUFAs in microalgal lipids are normally classified into two families, linolenic and α-linolenic acids. These two LC-PUFA act as energy-storing compounds and cell signaling molecules in the human cell membrane, providing significant health benefits. Additionally, some species of microalgae include the other two essential PUFAs, eicosapentaenoic and docosahexaenoic acids, which are crucial for human health. Microalgae oil offers a comparable amount of omega-3 to fish oil, but has the advantage of having less than three times the cholesterol level and no issue with pollutants or heavy metal contamination. One type of concentrated PUFA can be produced and stored by some species as complex polar lipids (such as phospholipids and glycolipids). The synthesis rate typically slows throughout stationary phase while the proportions of SFA and MUFA increase. PUFAs are typically produced at their highest levels during the early stationary phase.

8.2.10 Nanomaterial Synthesis

In recent decades, a great deal of research has been done on the utilization of microalgae for nanoparticles. Because of its ability to withstand high concentrations of heavy metals, *Chlorella sp.* is the species that is most frequently utilized to create nanoparticles. Because microalgae are non-virulent and most species are non-toxic, using them as a miniature reaction vessel for the creation of nanoparticles appears more promising than using other microbes. The primary benefit of employing algae is its capacity to use nutrient-rich wastewater as a growing medium. Furthermore, employing a nanobiorefinery technique, microalgae can result in the synthesis of several products. Because they are affordable, environmentally safe, and simple to handle, cyanobacteria and microalgae constitute an effective choice for the biosynthesis of NPs. With the shift in the species under consideration, however, the biosynthesis of nanoparticles needs to be improved. Metal recovery and nanoparticle biosynthesis work together to remove heavy metals from the environment by accumulating heavy metals in algal cells as a result of interactions with algae through bioaccumulation. The ability of algae to withstand heavy metals is a technique for using microalgae as nanofactories. According to several reports (Abou-Shanab et al., 2014), algae can bioaccumulate heavy metals. The internal reduction of metal to nanoform is necessary for the intracellular synthesis of nanoparticles. As a result, algae have the capacity to carry out the simultaneous tasks of bioremediation and nanoparticle production. Utilizing intracellular and extracellular biosynthesis routes, cyanobacteria and microalgae exhibit significant potential for the synthesis of Ag-NPs.

One of the commercially available microalgae species, *Chlorella vulgaris*, was employed to create AgCl-NPs with particles that were about 9.8, 5.7 nm in diameter. The usage of extract made from the plant Bifurcaria bifurcata for the production of CuO-NPs (Copper Oxide nanoparticles) was also investigated by a team of researchers. The same study evaluated the synthetic nanoparticles' capacity to inhibit bacterial growth. It also makes a number of other high and low value-added chemicals, making it a promising option for nanoparticle biosynthesis. One of the essential properties of microalgae that has been the main tenet of study in the area of microalgae-based biorefinery is biofuel production. Utilizing bioderived nanoparticles from the same species or another microalgae species makes it possible to extract oil from microalgae. Even though in these situations biosynthesis still need improvement, choosing between batch and continuous culture is important. Another option is to employ residual biomass from lipid extraction for nanoparticle biosynthesis. There is a link between biofuels, nanoparticles, and algae-based heavy metal bioremediation.

Microalgae are a dependable source of biofuel, which can be produced with increased efficiency employing nanoparticle biosynthesis and utilization due to the microalgae's capacity for bioaccumulation. Fe_3O_4@silica core-shell nanoparticles are used, according to a report, to harvest microalgae for biofuel. In the same study, the same NPs were utilized as a catalyst to improve the production of biodiesel from microalgae. The synthesis of Ag-NPs utilizing an aqueous extract of *Enteromorpha compressa* as a reducing agent demonstrates the usefulness of macroalgae in this sector. Specific species are chosen after testing various microalgal species for the production of Ag-NPs.

The use of several enzymes and co-factors makes up the biosynthesis mechanism. The process may also involve a number of other non-enzymatic compounds, such as flavonoids and phospholipids. It is environmentally benign to use microalgae to bioreduce metal ions to nanoparticles. Most microorganisms' biosynthesis is controlled by the molecules nicotinamide adenine dinucleotide hydrogen (NADH) and nicotinamide adenine dinucleotide phosphate (NADP). The NADH reductase enzyme, which receives an electron from NAD^+, converts metal ions into their nano form. The enzyme is oxidized as a result, and metal ions are reduced to nanoparticles Metal ions must be taken in by cells where nanoparticle production takes place for intracellular nanoparticles to develop. Thylakoids are where intracellular synthesis normally takes place, and $NADP^+$ is also essential for this process. With the aid of the split water molecule's electrons, $NADP^+$ is transformed into NADPH. When a proton gradient forms across the membrane, the water separates and donates an electron to photosystem II (PSII). The effectiveness of NADH reductases in producing metallic nanoparticles from metallic ions has been the subject of extensive research. Through the complexes of PSII and an electron transport chain, the process takes place in the inner membrane of the mitochondria. In microalgae, it is a well-known mechanism for nanoparticle generation. The premise for intracellular nanoparticle manufacturing is the interaction of metal ions with microalgae and their uptake by intracellular organelles through metabolic pathways. The enzyme-dependent reduction process begins once the metal ions are trapped inside the cell and reduces the metal ions in the cytoplasm. After that, the metallic NPs in the cell's organelles begin to stabilize, develop, and accumulate, which starts the nucleation process. The interaction of metal ions with proteins and amino acid moieties, which serve as capping agents and have a strong affinity for metal ions, is what gives rise to the stability of produced nanoparticles. The presence of the NADH-dependent nitrate reductase enzyme during the biosynthesis process suggests that the entire process involves the cell's electron transport chain. Normally, nitrate is converted to nitrite by the NADH-dependent nitrate reductase enzyme. In particular, the liberated electron helps Ag-NPs (Ag_2^+ to Ag^0) produce metallic nanoparticles. However, more complex processes are needed for the creation of gold nanoparticles, which then reduce Au^+-sulfide to Au^0. As hydrogenase enzymes split the molecular hydrogen, releasing electrons to reduce Au^{3+} to Au^0, molecular hydrogen is essential for the reduction of Au^{3+}.

8.2.11 Microalgae in Biological Activity

8.2.11.1 Anticancer Activity

Free radicals cause the development of cancer cells, while normal cells go through stages of initiation, promotion, and advancement. The natural pigments like chlorophyll, carotenoid and their derivatives act as cancer chemopreventive agents. These substances also exhibit anti-mutagenic and antitumorigenic properties, which help to inhibit the growth of cancer cells. Additionally, antioxidant and anticancer properties in phycobiliprotein pigments were discovered. The anticancer activity of the pigments was assessed using a cytotoxicity assay, and it was discovered that it was higher in nitrogen-stressed conditions.

8.2.11.2 Anti-inflammatory Activity

Following physical damage or invasion by harmful bacteria and viruses, the immune system becomes activated, which results in inflammation, a complex physiological process. It is possible for dietary ingredients to treat chronic inflammatory illnesses since substances like PUFAs and pigments from microalgae have shown the ability to reduce inflammation. The use of algal extracts as a functional food ingredient to reduce inflammation may potentially be advantageous. Numerous researchers' in vivo experiments proved that compounds like carotenoids in microalgae have anti-inflammatory and antioxidant properties. This confirms that microalgae can effectively cure inflammatory conditions.

8.2.12 Anti-angiogenic Activity

Blood vessel formation, or angiogenesis, is crucial for the development and repair of organs. Under pathological circumstances, uncontrolled angiogenesis is likely to hasten the spread of illnesses. Angiogenesis inhibitors are therefore necessary since they provide better treatment with fewer adverse effects. The organic extract of green algae was reported to have anti-angiogenic and anti-proliferative activities. MTT assay was used to measure the viability reduction of cell cultures in the presence and absence of the extracts, and experimental models of neovascularization were employed to examine the organic extract from the microalgae. Additionally, the extract demonstrated positive chemopreventive action against carcinogenesis.

References

Abou-Shanab, R. A. I., El-Dalatony, M. M., EL-Sheekh, M. M., Ji, M.-K., Salama, E.- S., Kabra, A. N., & Jeon, B.-H. (2014). Cultivation of a new microalga, Micractinium reisseri, in municipal wastewater for nutrient removal, biomass, lipid, and fatty acid production. *Biotechnology and Bioprocess Engineering, 19*(3), 510–518.

Ananthi, V., Raja, R., Carvalho, I. S., Brindhadevi, K., Pugazhendhi, A., & Arun, A. (2021). A realistic scenario on microalgae based biodiesel production: Third generation biofuel. *Fuel, 284*, 118965.

Ansari, F. A., Nasr, M., Guldhe, A., Gupta, S. K., Rawat, I., & Bux, F. (2020). Techno-economic feasibility of algal aquaculture via fish and biodiesel production pathways: A commercial-scale application. *Science of the Total Environment, 704*, 135259.

Arunachalam Sivagurulingam, A. P., Sivanandi, P., Pandian, S., Arumugamurthi, S. S., & Sircar, A. (2019). Optimization and kinetic studies on biodiesel production from microalgae (Euglena sanguinea) using calcium methoxide as catalyst. *Energy Sources, Part A: Recovery, Utilization, and Environmental Effects, 41*(12), 1497–1507.

Banu, J. R., Preethi, Kavitha, S., Gunasekaran, M., & Kumar, G. (2020). Microalgae based biorefinery promoting circular bioeconomy-techno economic and life-cycle analysis. *Bioresource Technology, 302*, 122822.

Chia, S. R., Ong, H. C., Chew, K. W., Show, P. L., Phang, S.-M., Ling, T. C., Nagarajan, D., Lee, D.-J., & Chang, J.-S. (2018). Sustainable approaches for algae utilisation in bioenergy production. *Renewable Energy, 129*, 838–852.

Da Silva, R. P., Rocha-Santos, T. A. and Duarte, A. C. (2016). Supercritical fluid extraction of bioactive compounds. TrAC Trends in Analytical Chemistry, *76*, 40–51.

Endalew, A. K., Kiros, Y., & Zanzi, R. (2011a). Heterogeneous catalysis for biodiesel production from Jatropha curcas oil (JCO). *Energy, 36*(5), 2693–2700.

Endalew, A. K., Kiros, Y., & Zanzi, R. (2011b). Inorganic heterogeneous catalysts for biodiesel production from vegetable oils. *Biomass and Bioenergy, 35*(9), 3787–3809.

Fernández, F. G. A., Reis, A., Wijffels, R. H., Barbosa, M., Verdelho, V., & Llamas, B. (2021). The role of microalgae in the bioeconomy. *New Biotechnology, 61*, 99–107.

Galadima, A., & Muraza, O. (2014). Biodiesel production from algae by using heterogeneous catalysts: A critical review. *Energy, 78*, 72–83.

Ginting, M. S. A., Azizan, M. T., & Yusup, S. (2012). Alkaline in situ ethanolysis of Jatropha curcas. *Fuel, 93*, 82–85.

Goh, B. H. H., Ong, H. C., Cheah, M. Y., Chen, W.-H., Yu, K. L., & Mahlia, T. M. I. (2019). Sustainability of direct biodiesel synthesis from microalgae biomass: A critical review. *Renewable and Sustainable Energy Reviews, 107*, 59–74.

Guldhe, A., Moura, C. V. R., Singh, P., Rawat, I., Moura, E. M., Sharma, Y., & Bux, F. (2017). Conversion of microalgal lipids to biodiesel using chromium-aluminum mixed oxide as a heterogeneous solid acid catalyst. *Renewable Energy, 105*, 175–182.

Hidalgo, P., Ciudad, G., Schober, S., Mittelbach, M., & Navia, R. (2015). Biodiesel synthesis by direct transesterification of microalga Botryococcus braunii with continuous methanol reflux. *Bioresource Technology, 181*, 32–39.

Islam, M. A., Magnusson, M., Brown, R. J., Ayoko, G. A., Nabi, M. N., & Heimann, K. (2013). Microalgal species selection for biodiesel production based on fuel properties derived from fatty acid profiles. *Energies, 6*(11).

Karpagam, R., Jawaharraj, K., & Gnanam, R. (2021). Review on integrated biofuel production from microalgal biomass through the outset of transesterification route: a cascade approach for sustainable bioenergy. *Science of the Total Environment, 766*, 144236.

Kesserwan, F., Ahmad, M. N., Khalil, M., & El-Rassy, H. (2020) Hybrid CaO/Al_2O_3 aerogel as heterogeneous catalyst for biodiesel production. *Chemical Engineering Journal, 385*, 123834.

Muhammad, G., Alam, M. A., Mofijur, M., Jahirul, M. I., Lv, Y., Xiong, W., Ong, H. C., & Xu, J. (2021). Modern developmental aspects in the field of economical harvesting and biodiesel production from microalgae biomass. *Renewable and Sustainable Energy Reviews, 135*, 110209.

Nascimento, I. A., Marques, S. S. I., Cabanelas, I. T. D., Pereira, S. A., Druzian, J. I., de Souza, C. O., Vich, D. V., de Carvalho, G. C., & Nascimento, M. A. (2013). Screening microalgae strains for biodiesel production: Lipid productivity and estimation of fuel quality based on fatty acids profiles as selective criteria. *Bioenergy Research, 6*(1), 1–13.

Pandit, P. R., & Fulekar, M. H. (2019). Biodiesel production from Scenedesmus armatus using egg shell waste as nanocatalyst. *Materials Today: Proceedings, 10*, 75–86.

Qu, W., Show, P. L., Hasunuma, T., Ho, S.-H. (2020). Optimizing real swine wastewater treatment efficiency and carbohydrate productivity of newly microalga Chlamydomonas sp. QWY37 used for cell-displayed bioethanol production. Bioresour. Technol. 123072.

Raeisossadati, M., Moheimani, N. R., & Parlevliet, D. (2019). Luminescent solar concentrator panels for increasing the efficiency of mass microalgal production. *Renewable and Sustainable Energy Reviews, 101*, 47–59.

Rahman, N., Ramli, A., Jumbri, K., & Uemura, Y. (2020). Biodiesel production from N. oculata microalgae lipid in the presence of Bi_2O_3/ZrO_2 catalysts. *Waste and Biomass Valorization*, 11.

Raj, J. V. A., Bharathiraja, B., Vijayakumar, B., Arokiyaraj, S., Iyyappan, J., & Kumar, R. P. (2019). Biodiesel production from microalgae Nannochloropsis oculata using heterogeneous poly ethylene glycol (PEG) encapsulated $ZnOMn^{2+}$ nanocatalyst. *Bioresource Technology, 282*, 348–352.

Saravanan, A. P., Mathimani, T., Deviram, G., Rajendran, K., & Pugazhendhi, A. (2018). Biofuel policy in India: A review of policy barriers in sustainable marketing of biofuel. *Journal of Cleaner Production, 193*, 734–747.

Singh, R., Bux, F., & Sharma, Y. C. (2020). Optimization of biodiesel synthesis from microalgal (Spirulina platensis) oil by using a novel heterogeneous catalyst, β-strontium silicate (β-Sr_2SiO_4). *Fuel, 280*, 118312.

Singh, R., Kumar, A., & Sharma, Y. C. (2019). Biodiesel synthesis from microalgae (Anabaena PCC 7120) by using barium titanium oxide (Ba_2TiO_4) solid base catalyst. *Bioresource Technology, 287*, 121357.

Sivaramakrishnan, R., & Incharoensakdi, A. (2018). Microalgae as feedstock for biodiesel production under ultrasound treatment—A review. *Bioresource Technology, 250*, 877–887.

Skorupskaite, V., Makareviciene, V., & Gumbyte, M. (2016). Opportunities for simultaneous oil extraction and transesterification during biodiesel fuel production from microalgae: A review. *Fuel Processing Technology, 150*, 78–87.

Su, Y., Song, K., Zhang, P., Su, Y., Cheng, J., & Chen, X. (2017). Progress of microalgae biofuel's commercialization. *Renewable and Sustainable Energy Reviews, 74*, 402–411.

Sun, J., Xiong, X., Wang, M., Du, H., Li, J., Zhou, D., & Zuo, J. (2019). Microalgae biodiesel production in China: A preliminary economic analysis. *Renewable and Sustainable Energy Reviews, 104*, 296–306.

Sydney, E. B., da Silva, T. E., Tokarski, A., Novak, A. C., de Carvalho, J. C., Woiciecohwski, A. L., Larroche, C., & Soccol, C. R. (2011). Screening of microalgae with potential for biodiesel production and nutrient removal from treated domestic sewage. *Applied Energy, 88*(10), 3291–3294.

Torres, S., Acien, G., García-Cuadra, F., & Navia, R. (2017). Direct transesterification of microalgae biomass and biodiesel refining with vacuum distillation. *Algal Research, 28*, 30–38.

Tsolcha, O. N., Patrinou, V., Economou, C. N., Dourou, M., Aggelis, G., & Tekerlekopoulou, A. G. (2021). Utilization of biomass derived from cyanobacteria based agro-industrial wastewater treatment and raisin residue extract for bioethanol production. Water, *13*, 486.

Veillette, M., Giroir-Fendler, A., Faucheux, N., & Heitz, M. (2017). Esterification of free fatty acids with methanol to biodiesel using heterogeneous catalysts: From model acid oil to microalgae lipids. *Chemical Engineering Journal, 308*, 101–109.

Velasquez-Orta, S. B., Lee, J. G. M., & Harvey, A. P. (2013). Evaluation of FAME production from wet marine and freshwater microalgae by in situ transesterification. *Biochemical Engineering Journal, 76*, 83–89.

Wágner, D. S., Radovici, M., Smets, B. F., Angelidaki, I., Valverde-Pérez, B., & Plósz, ´ B. G. (2016). Harvesting microalgae using activated sludge can decrease polymer dosing and enhance methane production via co-digestion in a bacterial-microalgal process. Algal Res. 20.

Wilkanowicz, S. I., Hollingsworth, N. R., Saud, K., Kadiyala, U., & Larson, R. G. (2020). Immobilization of calcium oxide onto polyacrylonitrile (PAN) fibers as a heterogeneous catalyst for biodiesel production. *Fuel Processing Technology, 197*, 106214.

Yusoff, M. N. A. M., Zulkifli, N. W. M., Sukiman, N. L., Chyuan, O. H., Hassan, M. H., Hasnul, M. H., Zulkifli, M. S. A., Abbas, M. M., & Zakaria, M. Z. (2021). Sustainability of palm biodiesel in transportation: A review on biofuel standard, policy and international collaboration between Malaysia and Colombia. *Bioenergy Research, 14*(1), 43–60.

Chapter 9
Utilization of Microalgal Deoiled Biomass

Microalgae are considered as the potential candidate for the biodiesel and energy source, but its commercial stability needs to be explored (Ho et al., 2020). Though, after the extraction of lipid from microalgae, a major portion of DOB is left behind. The DOB has copious amounts of carbohydrate, protein, cellulose, and some minerals which can be utilized for various applications (Chaudry et al., 2015). Generally, DOB of microalgae has been explored for animal feed. Some studies reported that the DOB of microalgae has been converted to gas by anaerobic digestion, biochar production, and catalytic hydrothermal gasification (Delrue et al., 2013; Wang et al., 2015). Recent advancement reported that DOB was explored for cellulose-based nanomaterial, whereas in some studies, the DOB was recycled for the nutrient recovery to cultivate microalgae which can reduce the cost of the process. The synthesis of carbon-based nanomaterial from DOB was highly explored by researchers.

9.1 Chemical-Based Synthesis of Nanomaterials from DOB

CNF and cellulose nanocrystals are successfully synthesized from DOB of microalgae like *Cladophora rupestris, Gelidium elegans, Nannochlor Open pond is oceanica, Nannochlor Open pond is salina* (Chen et al., 2016; Xiang et al., 2016). The chemical process involves three steps, such as alkali treatment, bleaching, and acid hydrolysis, are multistep and preferred for the purification cellulose-based nanomaterial. Chemical treatment is required for the extraction of cellulose to remove the amorphous component as lignin and hemicellulose are chemically linked. The alkali treatment involves NaOH which permits the hydrolysis of ester bonds among lignin and hemicellulose (Vasistha et al., 2023). Hence, the hydrolysis allows the lowering of hemicellulose content, whereas the lignin part is not affected by alkali treatment. Further, bleaching and acid treatment are the main steps in the process as it helps in reduction of pigment and remove the lignin content present in the cellulose (Sucaldito & Camacho, 2017).

M. P. Rai and S. Vasistha, *Microalgae Biotechnology for Wastewater Treatment, Resource Recovery and Biofuels*, https://doi.org/10.1007/978-3-031-31674-6_9

El Achaby et al. (2018) utilized red algae biomass *Gelidium elegans,* for the synthesis of CNC. The produced CNC was having a diameter ranging from 5.1 to 9.1 nm. In addition, the synthesized CNC was used to produce polyvinyl alcohol composite which can be used as nanofillers. Whereas in another report, the CNF was produced from DOB of *Cladophora rupestris* through enzyme hydrolysis, glucan was used as enzyme. The produced CNF from microalgae having high crystallinity and better thermal stability (El Achaby et al., 2018). The produced CNF from *NannochlorOpen pond is oceanica* by TEMPO mediated oxidation and proved to have great mechanical strength which can be employed for various nanofillers application. The average tensile strength of CNF was 3–4 GPa. Hence, the above studies indicate that the DOB of microalgae can be utilized for sustainable and biocompatible nanomaterial (Lee et al., 2018).

9.2 Hydrothermal Synthesis

The hydrothermal process is a heterogeneous reaction that occurs in presence of water, above ambient temperature, and pressure. The HTC can convert various feedstock into the solid material with high carbon content (Zhao et al., 2019). It offers advantages like robustness, free from chemicals or enzymes, and ease to use. The process is exothermic and effective only in water. Literature reported that this technique proved to be an effective alternate for the synthesis of carbonaceous compounds. Various microalgae were explored for the synthesis of nanomaterials using HTC (Table 8.1).

The *Spirulina platensis* was explored to produce nitrogen doped carbon material by the process of HTC. The sample was kept in PTFE lined autoclave provided with temperature between 180 and 240 °C for 24 h. They found that in the presence of glucose, the HTC produced was spherical in shape, and the diameter of the particle was >2 μm in size (Falco et al., 2012). In another study, the protein was isolated along with nutrient recovery from *Spirulina platensis* and *Chlorella vulgaris* through HTC (Zhao et al., 2019). The previous literature discussed the use of raw microalgae and glucose for the synthesis of N doped activated carbon by the process of HTC at temperature of 180 °C for 24 h, and they were used as electrodes in capacitors which are environmentally friendly (Sevilla et al., 2014). The defatted microalgae *Aurantiochytrium limacinum* SR21 was explored for the recovery of phosphorus and nitrogen water soluble fraction using HTC treatment at the high temperature. The 100% phosphorus and nitrogen recovered at 250 °C for 10 min for recultivation of microalgae. They discussed that at higher temperature, the yield of the phosphorus and nitrogen decreased (Aida et al., 2017). Singh et al. (2019) demonstrated the green energy approach where fluorescent nitrogen phosphorus carbon dots were synthesized from *Dunaliella salina* followed by HTC. These fluorescent carbon dots were earlier obtained by various methods like chemical oxidation and laser ablation, hence synthesizing it from microalgae makes it a green approach. The synthesized fluorescent nitrogen phosphorus carbon dots have the ability for sensing the toxic environmental containment like Hg (II) and Cr (IV) metal ions. Furthermore, Lu et al. (2015) used wet algae biomass of *NannochlorOpen pond is* sp. (15 wt %) slurry at 180–200 °C and 15–30 min to produce hydrochar which contain 85% of the FAMEs and 75% of the EPA. The recovery rate of 74% was achieved for fatty acids using ethanol. The study describes the promising method for the extraction of fatty acid by the HTC process and can be used for various applications in biofuel or nutraceutical (Table 9.1).

Table 9.1 Hydrothermal-based synthesis using microalgae

Source	Method	Solvent	Time (Hrs)	Temp (°C)	Material	References
Dunaliella salina	Hydrothermal synthesis	Double distilled water (DDW)	6	200	NP carbon dot	Singh et al. (2019)
Spirulina platensis	Hydrothermal carbonization	Water + glucose	24	180, 200, 220, 240	Nitrogen doped hydrothermal carbons	Falco et al. (2012)
Spirulina platensis	Hydrothermal co-carbonization	Glucose	24	180	N doped highly microporous carbon material	Sevilla et al. (2014)
Cladophora rupestris	Hydrothermal carbonization	Water	6	220	Carbon nano dots	Ana and Camacho (2019)
Gelidium amansii	Hydrothermal conversion	Water	2, 4, 6, 8, 10, 12	170	Cubic carbon quantum dots	Dou et al. (2018)
Sargassum horneri	Hydrothermal conversion	Water and citric acid	2, 4, 8, 16	180–210	Hydrochar	Zeng et al. (2018)
Nannochloropsis sp	Hydrothermal carbonization	Water	15, 30	180, 200, 220	Solid hydrochar	Lu et al. (2015)
Chlamydomon as reinhardtii	Hydrothermal carbonization	Oxalic acid & water	2	203	Algal char	Heilmann et al. (2010)
Nannochloropsis oculata	Hydrothermal carbonization	Water	15, 30, 45	215	Hydrochar	Levine et al. (2013)

(continued)

Table 9.1 (continued)

Source	Method	Solvent	Time (Hrs)	Temp (°C)	Material	References
Dunaliella salina	Hydrothermal method	Water	5	200	Nitrogen and sulfur co-doped carbon dots	Li et al. (2019)
Aurantiochytrium	Hydrothermal synthesis	Water	4	400	Extraction of nitrogen and phosphorous water soluble	Aida et al. (2017)
Chlamydomon as reinhardtii	Hydrothermal synthesis	Water	2	200	Char	Heilmann et al. (2011)
Spirulina platensis and Chlorella vulgaris	Hydrothermal synthesis	Water	1	190	Hydrochar	Zhao et al. (2019)

References

Aida, T. M., Maruta, R., Tanabe, Y., Oshima, M., Nonaka, T., Kujiraoka, H., Kumagai, Y., Ota, M., Suzuki, I., Watanabe, M. M., Inomata, H., & Smith, R. L. (2017). Nutrient recycle from defatted microalgae (Aurantiochytrium) with hydrothermal treatment for microalgae cultivation. *Bioresource Technology, 228*, 186–192.

Ana, J. C. S., & Camacho, D. H. (2019). Influence of precursor size in the hydrothermal synthesis of cellulose-based carbon nanodots and its application towards solar cell sensitization. *Materials Chemistry and Physics, 228*, 187–193.

Chaudry, S., Bahri, P. A., & Moheimani, N. R. (2015). Pathways of processing of wet microalgae for liquid fuel production: A critical review. *Renewable and Sustainable Energy Reviews, 52*, 1240–1250.

Chen, Y. W., Lee, H. V., Juan, J. C., & Phang, S.-M. (2016). Production of new cellulose nanomaterial from red algae marine biomass Gelidium elegans. *Carbohydrate Polymers, 151*, 1210–1219.

Delrue, F., Li-Beisson, Y., Setier, P.-A., Sahut, C., Roubaud, A., Froment, A.-K., & Peltier, G. (2013). Comparison of various microalgae liquid biofuel production pathways based on energetic, economic and environmental criteria. *Bioresource Technology, 136*, 205–212.

Dou, D., Duan, J., Zhao, Y., He, B., & Tang, Q. (2018). Cubic carbon quantum dots for light-harvesters in mesoscopic solar cells. *Electrochimica Acta, 275*, 275–280.

El Achaby, M., Kassab, Z., Aboulkas, A., Gaillard, C., & Barakat, A. (2018). Reuse of red algae waste for the production of cellulose nanocrystals and its application in polymer nanocomposites. *International Journal of Biological Macromolecules, 106*, 681–691.

Falco, C., Sevilla, M., White, R. J., Rothe, R., & Titirici, M.-M. (2012). Renewable nitrogen-doped hydrothermal carbons derived from microalgae. *Chemsuschem, 5*(9), 1834–1840.

Heilmann, S. M., Davis, H. T., Jader, L. R., Lefebvre, P. A., Sadowsky, M. J., Schendel, F. J., von Keitz, M. G., & Valentas, K. J. (2010). Hydrothermal carbonization of microalgae. *Biomass and Bioenergy, 34*(6), 875–882.

Heilmann, S., Jader, L., Harned, L., Sadowsky, M., Schendel, F., Lefebvre, P., Von Keitz, M., & Valentas, K. (2011). Hydrothermal carbonization of microalgae II. Fatty acid, char, and algal nutrient products. *Applied Energy, 88*, 3286–3290.

Ho, S.-H., Zhang, C., Tao, F., Zhang, C., & Chen, W.-H. (2020). Microalgal torrefaction for solid biofuel production. *Trends in Biotechnology, 38*(9), 1023–1033.

Lee, H.-R., Kim, K., Mun, S. C., Chang, Y. K., & Choi, S. Q. (2018). A new method to produce cellulose nanofibrils from microalgae and the measurement of their mechanical strength. *Carbohydrate Polymers, 180*, 276–285.

Levine, R., Sierra, C., Hockstad, R., Obeid, W., Hatcher, P., & Savage, P. (2013). The use of hydrothermal carbonization to recycle nutrients in algal biofuel production. *Environmental Progress and Sustainable Energy, 32.*

Li, Y., Liu, F., Cai, J., Huang, X., Lin, L., Lin, Y., Yang, H., & Li, S. (2019). Nitrogen and sulfur co-doped carbon dots synthesis via one step hydrothermal carbonization of green alga and their multifunctional applications. *Microchemical Journal, 147*, 1038–1047.

Lu, Y., Levine, R. B., & Savage, P. E. (2015). Fatty acids for nutraceuticals and biofuels from hydrothermal carbonization of microalgae. *Industrial and Engineering Chemistry Research, 54*(16), 4066–4071.

Sevilla, M., Gu, W., Falco, C., Titirici, M. M., Fuertes, A. B., & Yushin, G. (2014). Hydrothermal synthesis of microalgae-derived microporous carbons for electrochemical capacitors. *Journal of PowEr Sources, 267*, 26–32.

Singh, A. K., Singh, V. K., Singh, M., Singh, P., Khadim, S. R., Singh, U., Koch, B., Hasan, S. H., & Asthana, R. K. (2019). One pot hydrothermal synthesis of fluorescent NP-carbon dots derived from Dunaliella salina biomass and its application in on-off sensing of Hg (II), Cr (VI) and live cell imaging. *Journal of Photochemistry and Photobiology a: Chemistry, 376*, 63–72.

Sucaldito, M. R., & Camacho, D. H. (2017). Characteristics of unique HBr-hydrolyzed cellulose nanocrystals from freshwater green algae (Cladophora rupestris) and its reinforcement in starch-based film. *Carbohydrate Polymers, 169*, 315–323.

Vasistha, S., Balakrishnan, D., Manivannan, A., & Rai, M. P. (2023). Microalgae on distillery wastewater treatment for improved biodiesel production and cellulose nanofiber synthesis: A sustainable biorefinery approach. Chemosphere, *315*, 137666.

Wang, X., Zhao, B., Tang, X., & Yang, X. (2015). Comparison of direct and indirect pyrolysis of micro-algae Isochrysis. *Bioresource Technology, 179*, 58–62.

Xiang, Z., Gao, W., Chen, L., Lan, W., Zhu, J. Y., & Runge, T. (2016). A comparison of cellulose nanofibrils produced from Cladophora glomerata algae and bleached eucalyptus pulp. *Cellulose, 23*(1), 493–503.

Zeng, G., Lou, S., Ying, H., Wu, X., Dou, X., Ai, N., & Wang, J. (2018). Preparation of microporous carbon from Sargassum horneri by hydrothermal carbonization and KOH activation for CO_2 capture. *Journal of Chemistry, 2018*, 4319149.

Zhao, X., Stökle, K., Becker, G. C., Zimmermann, M., & Kruse, A. (2019). Hydrothermal carbonization of Spirulina platensis and Chlorella vulgaris combined with protein isolation and struvite production. *Bioresource Technology Reports, 6*, 159–167.

Challenges and Future Directions

The implementation of the biorefinery idea for microalgae production using industrial effluent is fraught with difficulties. Because one extraction method may affect the quality and quantity of other extraction products, multiproduct cascade extraction systems should be carefully followed. Acid treatment is far more harmful to proteins than sugar in terms of biomolecule structural degradation, protein extraction via alkali treatment (sodium hydroxide) should be carried out before sugar extraction via acid treatment. In order to obtain the highest quality biorefinery products, careful revisions must be made to the selection of the right extraction method and the extraction sequence of biomolecules from microalgae. Most of the microalgae biorefineries are not cost effective because of the unexplored new value-added multiple products from microalgal biomass grown on wastewater. Therefore, it is important to investigate new value-added biomaterials such as emulsifiers for bioplastic manufacture, carbon-dot, quantum dot, biomedical patches for wounds, or goods for other highly valuable medical uses. Due to the vast diversity of nutrients and their strength in a different kind of IWWs, they require multiple pretreatments to make a nutrient balance medium. In these situations, research should concentrate on using IWW cocktails that have been optimized from a variety of sources as a single, well-balanced nutritional media for microalgae culture. Depending on the worldwide region's temperature and sunlight availability, IWW valuation and microalgae production techniques may differ. Therefore, cold nations (those with low temperatures and little sunlight, like those in Europe) should concentrate on photobioreactor-based microalgae cultivation in order to produce high value-added products from the microalgae biomass cultivated on IWW. Southern Asia, the Middle East, and other hot regions should concentrate on raceway pond microalgae growing on IWW with the goal of numerous biorefineries-based biomass valorization. Therefore, a final route of future investigation would be estimating life cycle analysis of static microalgae wastewater treatment process to provide insight in overall sustainability.

M. P. Rai and S. Vasistha, *Microalgae Biotechnology for Wastewater Treatment, Resource Recovery and Biofuels*, https://doi.org/10.1007/978-3-031-31674-6

Bibliography

Aguirre, A. M., Bassi, A., & Saxena, P. (2013). Engineering challenges in biodiesel production from microalgae. *Critical Reviews in Biotechnology, 33*, 293.

Almutairi, A. W., El-Sayed, A. E. K. B., & Reda, M. M. (2020). Combined effect of salinity and pH on lipid content and fatty acid composition of Tisochrysis lutea. *Saudi Journal of Biological Sciences, 27*(12), 3553–3558.

Arora, N., & Philippidis, G. P. (2021). Insights into the physiology of Chlorella vulgaris cultivated in sweet sorghum bagasse hydrolysate for sustainable algal biomass and lipid production. *Scientific Reports, 11*(1), 6779.

Bajpai, P. (2019). Characteristics of algae. In P. Bajpai (Ed.) *Third generation biofuels* (S. 11–15). Springer Singapore.

Bi, Z., & He, B. (2020). Biodiesel from microalgae. In E. Jacob-Lopes, M. M. Maroneze, M. I. Queiroz, & L. Q. B. T.-H. of M.-B. P., & P. Zepka (Eds.) (S. 329–371). Academic Press.

Brindhadevi, K., Mathimani, T., Rene, E. R., Shanmugam, S., Chi, N. T. L., & Pugazhendhi, A. (2021). Impact of cultivation conditions on the biomass and lipid in microalgae with an emphasis on biodiesel. *Fuel, 284*, 119058.

Chang, H.-X., Fu, Q., Huang, Y., Xia, A., Liao, Q., Zhu, X., et al. (2016). An annular photobioreactor with ion-exchange-membrane for non-touch microalgae cultivation with wastewater. *Bioresource Technology, 219*, 668–676.

Ciurli, A., Di Baccio, D., Scartazza, A., Grifoni, M., Pezzarossa, B., Chiellini, C., Mariotti, L., & Pardossi, A. (2021). Influence of zinc and manganese enrichments on growth, biosorption and photosynthetic efficiency of *Chlorella sp. Environmental Science and Pollution Research, 28*(7), 8539–8555.

Corredor, L., Barnhart, E. P., Parker, A. E., Gerlach, R., & Fields, M. W. (2021). Effect of temperature, nitrate concentration, pH and bicarbonate addition on biomass and lipid accumulation in the sporulating green alga PW95. *Algal Research, 53*, 102148.

De la Cruz, A. A., Chernoff, N., Sinclair, J. L., Hill, D., Diggs, D. L., & Lynch, A. T. (2020). Introduction to cyanobacteria and cyanotoxins. In *Water treatment for purification from cyanobacteria and cyanotoxins* (S. 1–35).

Demoulin, C. F., Lara, Y. J., Cornet, L., François, C., Baurain, D., Wilmotte, A., & Javaux, E. J. (2019). Cyanobacteria evolution: Insight from the fossil record. *Free Radical Biology and Medicine, 140*, 206–223.

Erickson, E., Wakao, S., & Niyogi, K. K. (2015). Light stress and photoprotection in Chlamydomonas reinhardtii. *The Plant Journal, 82*(3), 449–465.

Fang, L., Leliaert, F., Zhang, Z.-H., Penny, D., & Zhong, B.-J. (2017). Evolution of the chlorophyta: Insights from chloroplast phylogenomic analyses. *Journal of Systematics and Evolution, 55*(4), 322–333.

M. P. Rai and S. Vasistha, *Microalgae Biotechnology for Wastewater Treatment, Resource Recovery and Biofuels*, https://doi.org/10.1007/978-3-031-31674-6

González-Trujillo, J. D., Pedraza-Garzón, E., Donato-Rondon, J. C., & Sabater, S. (2020). Ecoregional characteristics drive the distribution patterns of neotropical stream diatoms. *Journal of Phycology, 56*(4), 1053–1065.

Gunderson, J. H., Elwood, H., Ingold, A., Kindle, K., & Sogin, M. L. (1987). Phylogenetic relationships between chlorophytes, chrysophytes, and oomycetes. *Proceedings of the National Academy of Sciences, 84*(16), 5823 LP–5827.

Harwood, J. L., & Guschina, I. A. (2009). The versatility of algae and their lipid metabolism. *Biochimie, 91*(6), 679–684.

Heimann, K., & Huerlimann, R. (2015). *Microalgal classification: Major classes and genera of commercial microalgal species*. In S.-K. B. T.-H. of M. M. Kim (Eds.) (S. 25–41). Academic Press.

Hu, Q., Kurano, N., Kawachi, M., Iwasaki, I., & Miyachi, S. (1998). Ultrahigh-cell-density culture of a marine green alga Chlorococcum littorale in a flat-plate photobioreactor. *Applied Microbiology and Biotechnology, 49*(6), 655–662.

Ji, B., Zhu, L., Wang, S., & Liu, Y. (2021). Temperature-effect on the performance of non-aerated microalgal-bacterial granular sludge process in municipal wastewater treatment. *Journal of Environmental Management, 282*, 111955.

Kamalanathan, M., Chaisutyakorn, P., Gleadow, R., & Beardall, J. (2018). A comparison of photoautotrophic, heterotrophic, and mixotrophic growth for biomass production by the green alga Scenedesmus sp. (Chlorophyceae). *Phycologia, 57*(3), 309–317.

Karimi, M. (2016). Exergy-based optimization of direct conversion of microalgae biomass to biodiesel. *Journal of Cleaner Production, 141.*

Khan, M. I., Shin, J. H., & Kim, J. D. (2018). The promising future of microalgae: current status, challenges, and optimization of a sustainable and renewable industry for biofuels, feed, and other products. *Microbial Cell Factories, 17*(1), 36.

Kumar, A., Ergas, S., Yuan, X., Sahu, A., Zhang, Q., Dewulf, J., Malcata, F. X., van Langenhove, H. (2010). *Enhanced CO(2) fixation and biofuel production via microalgae: Recent developments and future.*

Li, X., Manuel, J., Slavens, S., Crunkleton, D. W., & Johannes, T. W. (2021). Interactive effects of light quality and culturing temperature on algal cell size, biomass doubling time, protein content, and carbohydrate content. *Applied Microbiology and Biotechnology, 105*(2), 587–597.

Lima, S., Schulze, P. S. C., Schüler, L. M., Rautenberger, R., Morales-Sánchez, D., Santos, T. F., Pereira, H., Varela, J. C. S., Scargiali, F., Wijffels, R. H., & Kiron, V. (2021). Flashing light emitting diodes (LEDs) induce proteins, polyunsaturated fatty acids and pigments in three microalgae. *Journal of Biotechnology, 325*, 15–24.

Lu, W., Liu, S., Lin, Z., & Lin, M. (2021). Enhanced microalgae growth for biodiesel production and nutrients removal in raw swine wastewater by carbon sources supplementation. *Waste and Biomass Valorization, 12*(4), 1991–1999.

Malla, F. A., Bandh, S. A., Wani, S. A., Hoang, A. T., & Sofi, N. A. (2022). Biofuels: Potential alternatives to fossil fuels. In: Bandh, S. A., & Malla, F. A. (Eds.), *Biofuels in circular economy*. Springer, Singapore. https://doi.org/10.1007/978-981-19-5837-3_1

Mamtani, K., Shahbaz, K., & Farid, M. M. (2021). Deep eutectic solvents–versatile chemicals in biodiesel production. *Fuel, 295*, 120604.

Mantzorou, A., & Ververidis, F. (2019). Microalgal biofilms: a further step over current microalgal cultivation techniques. *Science of the Total Environment, 651*, 3187–3201.

Marella, T. K., López-Pacheco, I. Y., Parra-Saldívar, R., Dixit, S., & Tiwari, A. (2020). Wealth from waste: Diatoms as tools for phycoremediation of wastewater and for obtaining value from the biomass. *Science of the Total Environment, 724*, 137960.

Matsumoto, T., Shinozaki, F., Chikuni, T., Yabuki, A., Takishita, K., Kawachi, M., Nakayama, T., Inouye, I., Hashimoto, T., & Inagaki, Y. (2011). Green-colored plastids in the dinoflagellate genus lepidodinium are of core chlorophyte origin. *Protist, 162*(2), 268–276.

Metsoviti, M. N., Papapolymerou, G., Karapanagiotidis, I. T., & Katsoulas, N. (2020b). Effect of light intensity and quality on growth rate and composition of chlorella vulgaris. *Plants, 9*(1).

Metsoviti, M. N., Papapolymerou, G., Karapanagiotidis, I. T., & Katsoulas, N. (2020a). Effect of light intensity and quality on growth rate and composition of Chlorella vulgaris. *Plants, 9*(1), 1–17.

Mirhosseini, N., Davarnejad, R., Hallajisani, A., Cano-Europa, E., Tavakoli, O., Franco-Colín, M., & Blas-Valdivia, V. (2021). Research article: Cultivations of Arthrospira maxima (Spirulina) using ammonium sulfate and sodium nitrate as an alternative nitrogen sources. *IFRO, 20*(2), 475–489.

Morales-Sánchez, D., Martinez-Rodriguez, O. A., & Martinez, A. (2017). Heterotrophic cultivation of microalgae: Production of metabolites of commercial interest. *Journal of Chemical Technology & Biotechnology, 92*(5), 925–936.

Oliveira, C. Y. B., D'Alessandro, E. B., Antoniosi Filho, N. R., Lopes, R. G., & Derner, R. B. (2021). Synergistic effect of growth conditions and organic carbon sources for improving biomass production and biodiesel quality by the microalga Choricystis minor var. minor. *Science of the Total Environment, 759*, 143476.

Oswald, W. J., & Golueke, C. G. (1960). Biological transformation of solar energy. In *Advances in applied microbiology* (Vol. 2, S. 223–262). Academic Press.

Penhaul Smith, J. K., Hughes, A. D., McEvoy, L., Thornton, B., & Day, J. G. (2021). The carbon partitioning of glucose and DIC in mixotrophic, heterotrophic and photoautotrophic cultures of Tetraselmis suecica. *Biotechnology Letters, 43*(3), 729–743.

Raven, J. A., Beardall, J., & Quigg, A. (2020). Light-driven oxygen consumption in the water-water cycles and photorespiration, and light stimulated mitochondrial respiration. In A. W. D. Larkum, A. R. Grossman, & J. A. Raven (Eds.), *Photosynthesis in algae: Biochemical and physiological mechanisms* (S. 161–178). Springer International Publishing.

Rohitha Thangam, K., Santhiya, A., Abinaya Sri, S. R., Mubarak Ali, D., Karthikumar, S., Shyam Kumar, R., Thajuddin, N., Soosai, M. R., Varalakshmi, P., Ganesh Moorthy, I., & Pugazhendhi, A. (2021). Bio-refinery approaches based concomitant microalgal biofuel production and wastewater treatment. *Science of the Total Environment,* 147267.

Sajjadi, B., Chen, W.-Y., Raman, A. A. A., & Ibrahim, S. (2018). Microalgae lipid and biomass for biofuel production: A comprehensive review on lipid enhancement strategies and their effects on fatty acid composition. *Renewable and Sustainable Energy Reviews, 97*, 200–232.

Sánchez-Baracaldo, P., & Cardona, T. (2020). On the origin of oxygenic photosynthesis and cyanobacteria. *New Phytologist, 225*(4), 1440–1446.

Singh, V., Yadav, V. K., & Mishra, V. (2020b). Nanotechnology: An application in biofuel production. In M. Srivastava, N. Srivastava, P. K. Mishra, & V. K. Gupta (Eds.), *Nanomaterials in biofuels research* (S. 143–160). Springer Singapore.

Singh, J., Jain, D., Agarwal, P., & Singh, R. P. (2020a). Auxin and cytokinin synergism augmenting biomass and lipid production in microalgae Desmodesmus sp. JS07. *Process Biochemistry, 95*, 223–234.

Slegers, P. M., Lösing, M. B., Wijffels, R. H., van Straten, G., & van Boxtel, A. J. B. (2013). Scenario evaluation of open pond microalgae production. *Algal Research, 2*(4), 358–368. https://doi.org/10.1016/j.algal.2013,05(001),pp.63

Sabóia-Morais, S. M. T. (2021). Co-exposure of iron oxide nanoparticles and glyphosate-based herbicide induces DNA damage and mutagenic effects in the guppy (Poecilia reticulata). *Environmental Toxicology and Pharmacology, 81*, 103521.

Syed Hosseini, N. S., Shang, H., Ross, G. M., & Scott, J. A. (2015). Microalgae cultivation in a novel top-lit gas-lift open bioreactor. *Bioresource Technology, 192*, 432–440.

Tan, Z. J., Yang, Z. Z., Yi, Y. J., Wang, H. Y., Zhou, W. L., Li, F. F., & Wang, C. Y. (2016). Extraction of oil from flaxseed (Linum usitatissimum L.) using enzymeassisted three-phase partitioning. *Applied Biochemistry and Biotechnology.*

Tandeau de Marsac, N. (2003). Phycobiliproteins and phycobilisomes: the early observations. *Photosynthesis Research, 76*(1), 193–205.

Watson, J., Swoboda, M., Aierzhati, A., Wang, T., Si, B., & Zhang, Y. (2021). Biocrude oil from algal bloom microalgae: A novel integration of biological and thermochemical techniques. *Environmental Science & Technology, 55*(3), 1973–1983.

Yaakob, M. A., Mohamed, R. M., Al-Gheethi, A., Aswathnarayana Gokare, R., & Ambati, R. R. (2021). Influence of nitrogen and phosphorus on microalgal growth, biomass, lipid, and fatty acid production: An overview. *Cells, 10*(2).

www.ingramcontent.com/pod-product-compliance
Ingram Content Group UK Ltd.
Pitfield, Milton Keynes, MK11 3LW, UK
UKHW022209051026
15624UKWH00001B/81
9783031316739